Openpit Mining

Red Alert is the deliberately urgent title for a new series of Earth Island texts dealing with specific issues on an ever-growing front of environmental concern.

The texts are short, from 15,000 to approximately 25,000 words, the simple assumption being that if the problem is as pertinent as is claimed, and the writer sufficiently sure and lucid in the handling of his material, he and his publisher should be capable of producing a work quickly, which the intelligent layman will read and absorb easily, and in certain ways, act upon.

Though brief, 'Red Alerts' are authoritative, comprehensively annotated when necessary, and include both index and bibliography. In the sense that any book must be of its time and place to be of real value, both their frames of reference and their data are contemporary and apposite. And in the sense that no book can ever be complete, they are calculated to lead both author and reader on to further examination of the subject.

First titles are *Nuclear Reactors* and *Openpit Mining*. Additional titles scheduled for 1973 are an investigation of The Bread Industry, Foulness, and Transport. Further titles may include The Oil Industry, Plastics Industry, Supersonic Flight, The Drug Industry. . . .

Sean F. Gallagher   Series Editor

Openpit Mining

Amory B. Lovins

No. 2 in a series of urgent environmental studies

Earth Island

First published in Great Britain 1973
Earth Island Limited
56 Doughty Street
London WC1N 2LS

Case Edition ISBN 0 85644 0302
Paper Edition ISBN 0 85644 020 5

[redacted]

Series cover design by Marisa Martelli

Text set in 10/12 pt. Monotype Baskerville, printed by letterpress, and bound in Great Britain at The Pitman Press, Bath

Contents

List of Tables

Acknowledgements

For their diverse and stimulating views on mineral resources, I am grateful to Mr William W. Behrens III (MIT), Mr David Brower (President, Friends of the Earth Inc), Mr Geoffrey Cox (Royal School of Mines), Professor Herman Daly (Louisiana State University), Dr Rustam Kothavala (Harvard University), Professor Thomas Lovering (Committee on Resources and Man, National Academy of Sciences/National Research Council), Messrs Walter C. Patterson and Graham Searle (both of Friends of the Earth Ltd), Mr Michael West (Editorial Director of *Mining Journal*), and Professor Carroll Wilson (MIT).

Applying these general views to specific mining problems would have been impossible without the files and insights painstakingly assembled by Messrs Simon Millar and Oliver Thorold, respectively staff geologist and staff barrister of Friends of the Earth Ltd. And explaining these problems would have been impossible without phrases and similes shamelessly recycled from many sources; Mr James Gardner deserves special mention for his vivid postage-stamp-on-a-football image of the earth's crust, in his Story of the Earth exhibition, Royal Geological Museum, South Kensington, London.

The wicked, it is written, shall fall into the pit which they have digged; so too the erroneous. But if I have fallen into error, the fault rests with me, not with those dedicated people within the mining industry who commented on the manuscript and helped unstintingly in so many other ways. I hope that these people will come to feel less lonely, so that they need no longer be anonymous within their profession.

—ABL

Preface

Openpit mining is one of the remarkable technologies that make us who are in the rich countries rich. So far it has been practised mainly in poor countries and in remote areas of large rich countries. We have reaped most of its benefits; others, out of our sight, have borne its often heavy costs.

Now the costs, too, are coming home. Mining companies, spurred by scarcity, are seeking to mine what is thine – in National Parks, for example. We have long averted our eyes from the technology that has made us so rich; belatedly, we are now forced by its new nearness to make urgent and perplexing ethical choices about it, and thus to discover just what our wealth costs. This book is designed to help you in those choices and in that discovery.

Mining companies do both good and bad. Yet even those that deserve criticism for particular tactics or decisions are acting more or less rationally within the accepted rules of the highly competitive resource-using game. Whether these rules are the right rules, whether they will on balance give you what you want, it is for you to judge: for ultimately the rules are made not by Ministers in Cabinet meetings, nor by directors in their boardrooms, but by you in your daily life.

—ABL

INTRODUCTION

On my desk are two fist-sized pieces of rock that tell a fascinating and important story. The first looks a bit like a plum pudding: it is liberally studded with beautiful cherry-sized crystals of the copper-bearing mineral chalcopyrite. This rock, roughly one-tenth copper, was taken years ago from one of the richest mines in Zambia – though nowadays, even there, you would have trouble finding such a sample.

The other rock has no iridescent knobs sticking out of it: it is just a dull grey lump. Only through a magnifying glass can you see that it too contains chalcopyrite – in tiny grains scattered so sparsely through the barren 'host' rock that you would have to sift through 200 tons of rock like this to get out one ton of copper. This rock comes from a newly opened mine typical of a fairly new breed: huge, highly mechanized openpit mines that can scoop out mountains like eggs, move tens of millions of tons of rock each year, and mill to a fine powder tens of tons of rock per minute so that the tiny grains of metal-rich minerals can be extracted. A single such mine in Britain, big enough to supply one-tenth of Britain's copper demand, would have to move more rock each year than is mined in the whole Zambian copper-belt.

Man did not resort by chance to larger mines and poorer ores: he was driven by an insatiable hunger for minerals. The amount of mining in the world is doubling about every fourteen years[1,2], more than twice as fast as the world population. Each man is thus spending the earth's mineral capital more than twice as fast as his father did. Nearly all the demand is in the rich countries: the USA alone has used more minerals in the past thirty years than the whole world used in all time before[3]. And as the smaller, more easily mined deposits of rich ores are thus used up, the character of the mines must change, generally towards the large modern kind that this book is about.

Openpit mining has a dramatic impact on land and is increasingly popular, but it is only one of many ways of extracting minerals from the earth. The general families of mining methods include:

- underground mining, where only a shaft or tunnel breaks the surface;
- dredging or scooping up loose *placer* deposits such as gold dust or tin, laid down by running water in river-beds, on beaches, or on the seabed[4];
- pumping a liquid or gas up to the surface, as with petroleum (soluble minerals such as potash can be *solution mined* – dissolved underground and pumped up as a solution);
- extracting salts dissolved in seawater or brine;
- recovering substances from slag and other by-products of smelting and refining, from flue gases, from coal ash, etc.;
- surface mining, of which openpit mining is one kind.

Surface mining, which contributes[3] about 95 per cent of all US non-fuel minerals, takes many forms ranging upwards in size from the common sand- or gravel-pit. Perhaps the best-known surface method is the *strip mining* or *opencast mining* of shallow coal seams and other thin layers of mineral lying near the surface. The mineral can often be extracted by progressively turning over areas of soil, rather like a wave in slow motion, so that *overburden* – waste soil and rock covering the mineral – is used to fill the space from which mineral has just been extracted. This method disturbs a large area, but if it is done carefully, land that is neither steep nor fragile can often be restored to a useful and even fairly fertile state afterwards – at a cost of thousands of pounds sterling per hectare (2.47 acres = 1 hectare, very roughly twice the size of a football field).

On the other hand, a body of ore that is thicker or more deeply buried, or that dips sharply downward, may be better mined by digging a broad open pit around and including it. Such a pit is often so deep – from a few hundred to more than a thousand feet (100–500 metres) – that it cannot be filled in again. Since it is made more by digging downwards than by scraping sideways, it is not generally suited to continuous or progressive refilling. Though both this deep-pit method and the shallower surface-stripping are often loosely called *opencast*, I shall avoid confusion by calling the deeper method *openpit*. It may occupy a smaller area than stripping would, but its damage is more lasting.

Because every orebody is unique, there are no general rules of mining, and a single mine may commonly use several different methods at different places or times; it may start as a stripping operation and end as an open pit, or be both at once in different parts, or start as a pit and then continue underground from inside the pit. And it is sometimes hard to decide the category of a mine: there are many intermediate cases.

Mines for different minerals are often built on completely different geological principles. Opencast coal-mining, for example, is not at all like openpit copper-mining, nor underground tin-mining like underground uranium-mining: the geology is different for each. Because there are far too many kinds of underground, solution, and placer mining to discuss their distinctive problems in such a short space, this Red Alert text will not discuss in any detail those minerals that do not normally occur in a form suitable for openpit mining.

Openpit mines themselves also vary greatly. One major division is between quarrying, where inherently valuable rock is mostly cut into blocks and sold with little treatment, and openpit metal-mining, where much of the rock is ground to a powder and then discarded as tailings after the metal-bearing mineral grains have been removed. Such a mine must move more rock and produce more waste for a given yield of saleable product. Within and between these two classes of openpit mining – which both produce large holes in the ground – there are dozens of specialized kinds of mining. The problems of openpit mining for coal, borax, clay, gypsum, potash, asbestos, fluorspar, sand, gravel, limestone, granite, phosphate rock, slate, and so on are so varied that a work of this length must concentrate on the one kind of openpit mining most important to the general public (especially in future): openpit metal-mining. The general techniques and problems of this sort of mining are indirectly relevant to mining for non-metallic minerals too; so with the warning that there are also important differences, I shall ignore from here on all of the specialized topics not directly related to metal-mining.

This emphasis on openpit methods reflects recent and probable future trends in world metal-mining: concentrated ores best worked selectively by hand are giving way to disseminated ores workable only by mountain-moving. In 1836, when Britain produced 45 per cent of the world's copper[4], the underground mines of Cornwall were working ore containing 8.3 per cent copper[5]. In 1936, when the USA had attained its present status as the world's largest copper-producer, only 55 per cent of US copper came from underground mines (the rest was openpit), and the average ore contained 1.5 per cent copper[4,6]. In 1968, when the average grade had fallen to 0.60 per cent copper[3], nearly three times as much copper was produced in the US – 83 per cent of it by openpit mining[3].

Many metals show a similar trend; indeed, a few have been mined only by openpit methods because their geology is unsuited to anything else. As rising demand depletes rich ores, openpit mining will become more popular and profitable. For all the metals for which it can be done, we shall be seeing it done more often, both within and outside traditional mining areas. It is mainly within these areas – including the beautiful British highlands, much of them in National Parks – that more than thirty mining companies are now pursuing more than 100 prospects, many of low grade, in a new British mineral rush like that in many other densely populated countries.

Except for differences due to climate, the general form of openpit metal-mining is much the same round the world, so that little generality is lost by concentrating on mines and prospects in Britain, Western Europe, and North America. This book is not a world survey, and therefore has some regional bias. Discussions that must be specific use UK examples; for resources, on the other hand, the perspective is global. US mining data appear often, simply because the thorough statistics published by the US Bureau of Mines have no equally comprehensive and accessible counterpart in other countries.

A note on units: 'ton' always means 'metric ton' (1000 kilograms or 2205 pounds), *not* 'short ton' (1 ston = 2000 pounds) or 'long ton' (1 lton = 2240 pounds). 'Million' means '1,000,000', billion '1,000,000,000'. Many units are given in both English and metric (original unit first, derived unit second): the mining industry uses mainly metric. Precious metals are commonly measured by the troy ounce (32.15 to the kilogram, 14.58 to the pound avoirdupois), but in this book they are all in grams (exactly 1 million to the [metric] ton). 'Metre' (39.36 inches) is abbreviated 'm'; 'square metre', 'm^2'; 'cubic metre', 'm^3'; 'foot' ('ft') and 'yard' ('yd') likewise; 'centimetre' (0.01 m or 2.54 inches) is 'cm', 'kilogram' (0.001 ton or 2.2 pounds) is 'kg', and 'gram' (0.001 kg or 0.0353 ounce avoirdupois) is 'g'. Areas are measured by the m^2, hectare (10,000 m^2), acre (4046.8 m^2 = 43,650 ft^2), and square mile (640 acres or 2.59 million m^2).

Geology and mining engineering, like the other disciplines (economics, sociology, law, etc.) into which this book strays, are complex and highly technical subjects. Yet in essence they are simple enough so that anyone who needs to know the nature and implications of mining can be given an adequate working knowledge without using jargon or sacrificing technical honesty. Readers who wish to supplement common-sense descriptions with full scientific detail will find initial guidance in the references to each chapter.

This book is written primarily for that lay group who will need to understand and query expert judgements, to read between professionally drafted lines, to weigh the desires of mining companies in the scales of social purpose, and to ask three new questions: What does it cost our children? What does it cost the earth? What if we didn't do it?

Notes

1. *Statistical Yearbook 1971*, United Nations, 1972.
2. Study of Critical Environmental Problems: *Man's Impact on the Global Environment*, MIT Press, 1970.
3. *Mineral Facts and Problems*, US Bureau of Mines, Bulletin 650, 1970.
4. US National Academy of Sciences/National Research Council, Committee on Resources and Man (Preston Cloud, Chairman): *Resources and Man*, W. H. Freeman (San Francisco), 1969.
5. H. T. de la Beche: *Report on the Geology of Cornwall, Devon, and West Somerset*, Longman, Orme, Brown, Green, and Longmans (London) for HMSO, 1839.
6. E. Pfleider: *Surface Mining*, Amer. Inst. Min. Metall. Petrol. Engineers (New York), 1968.

1

METALS

The typewriter I am now using probably contains[1] Jamaican or Surinam aluminium, Swedish iron, Czech magnesium, Gabonese manganese, Rhodesian chromium, Soviet vanadium, Peruvian zinc, New Caledonian nickel, Chilean copper, Malaysian tin, Nigerian columbium, Zairean (Congolese) cobalt, Yugoslav lead, Canadian molybdenum, French arsenic, Brazilian tantalum, South African antimony, Mexican silver, and traces of other well-travelled metals. The enamel may contain Norwegian titanium; the plastic is made of Middle Eastern oil (cracked with American rare-earth catalysts) and of chlorine (extracted with Spanish mercury); the foundry sand came from an Australian beach; the machine tools used Chinese tungsten; the coal came from the Ruhr; the end product consumes, some might say, too many Scandinavian spruces.

I give this list not to emphasize the remarkable and perhaps precarious interdependence of world industry, but to bring home an easily forgotten fact: that every bit of my typewriter, somewhere, sometime, was a mineral in the ground. Through smelting and refining processes, fuelled by burning coal and petroleum, the pure metals have been liberated from metal-bearing minerals. These in turn have been discovered, mined, and processed as metallic ores. To understand these processes, we must begin with the metals themselves, and then study how metals are bound in minerals, how minerals occur in ores, and how ores are distributed in the earth's crust.

Not all metals are mined in their own right. Some for which there is less demand are produced mainly or wholly as *by-products* of others that are the main object of mining. Occasionally metals are *co-products* – they are mined together where it might not pay to mine just for one.

The metals that are the primary objects of modern mining are listed in Table 1, each with its chemical symbol as one of the ninety-two naturally occurring chemical elements. Each metal is classified[1], according to the US Bureau of Mines system, as a nuclear reactor fuel – coded 'R' in the table – or as a ferrous ('F'), non-ferrous ('N'), or minor ('M') metal. (Minor metals have special uses but are seldom alloyed. In the USBM system, ferrous metals are those commonly used in steels; a chemist might not classify metals in quite the same way.) Some main uses[1] are also given for each metal, along with its *clarke*[1,2] – its approximate average abundance in the earth's crust, expressed by Table 1 in parts per million (ppm) by weight. This is the same as grams per ton; to convert to a percentage, just move the decimal point four places to the left.

Table 2 lists the main metals produced as by-products or co-products of the *primary metals* in Table 1, showing the usual source[1] of each. These *secondary metals* may be concentrated from their host ores at the mine, separated during smelting or refining, or recovered from wastes of those processes.

TABLE 1

PRIMARY METALS

Type	*Metal*§	*Symbol*	*Clarke*	*Main Uses*
N	Aluminium	Al	81,300	Structural alloys, electrics, packaging
F	Iron	Fe	50,600	Cast iron, all steels, magnets
N	Magnesium	Mg	20,900	Firebrick, aircraft alloys, pigments
N	Titanium	Ti	4400	Pigments, paper, plastics, aircraft alloys
F	Manganese	Mn	1000	Nearly all iron and steel
M	Strontium	Sr	300	Pyrotechnics, Zn refining, grease, ceramics
M	Barium	Ba	250	Drilling oil/gas wells, glass
N	Zirconium	Zr	220	Sand-casting, ceramics, nuclear reactors
F	Chromium	Cr	200	Alloys, plating, pigments, tanning
F	Vanadium	V	150	Steels, catalysts
N	Zinc	Zn	132	Brass, galvanizing, diecastings, rubber, paints
N	Rare-earth metals	*	total ~121	Oil refining, glass, ceramics, electronics
F	Nickel	Ni	80	Alloys, catalysts, batteries, magnets
M	Lithium	Li	65	Ceramics, glass, greases, welding
N	Copper	Cu	55	Electrics, brass, bronze, heat-transfer tubing
N	Tin	Sn	40	Cans, bronzes and other alloys, solders
F	Columbium (niobium)	Cb (Nb)	24	Special alloys, magnets, pigments, ceramics
N	Lead	Pb	16	Batteries, petrol, paints, chemicals, alloys
R	Thorium	Th	~ 12	Nuclear fuel, alloys, catalysts, gas mantles
F	Molybdenum	Mo	5	Steels, alloys, chemicals, pigments, lubricants
R	Uranium	U	2.5	Nuclear fuel and weapons
F	Tantalum	Ta	2	Electronics, alloys
F	Tungsten (wolfram)	W	< 2	Superhard alloys, lamp filaments
N	Antimony	Sb	1	Batteries, type metal, special alloys
N	Mercury	Hg	< 0.4	Batteries, lamps, paints, instruments, chlorine
N	Silver	Ag	0.1	Silverware, photography, coins, electronics
N	Platinum group	†	total ~0.019	Catalysts, oil refining, instruments, anti-pollution devices
N	Gold	Au	0.005	Bullion, jewellery, electronics

* The rare-earth metals, with their symbols and clarkes, are: cerium (Ce, 46), neodymium (Nd, 24), lanthanum (La, 18), samarium (Sm, 6.5), gadolinium (Gd, 6.4), praseodymium (Pr, 5.5), dysprosium (Dy, 4.5), ytterbium (Yb, 2.7), erbium (Er, 2.5), holmium (Ho, 1.2), europium (Eu, 1.1), terbium (Tb, 0.9), lutetium (Lu, 0.8), thulium (Tm, 0.2), and promethium (Pm, essentially 0).

† The platinum-group metals, with their symbols and clarkes, are: palladium (Pd, 0.01), platinum (Pt, 0.005), rhodium (Rh, 0.001), iridium (Ir, 0.001), osmium (Os, ~ 0.001), and ruthenium (Ru, ~ 0.001).

§ Because they have no uses comparable to those of the metals listed in this table, the reactive alkali metals sodium (Na; clarke = 28,300), potassium (K; 25,900), and calcium (Ca; 36,300) have been omitted. They are normally obtained from salt, potash, and limestone respectively, but are rarely used as metals or in alloys.

TABLE 2

SECONDARY METALS

Type	*Metal*	*Symbol*	*Clarke*	*Main Uses*	*Main Sources*
N	Rubidium	Rb	310	Space, electronic research	by Cs
N	Yttrium	Y	28	Electronics, alloys, ceramics	by U, co rare earths
F	Cobalt	Co	23	Alloys, magnets, pigments, ceramics	by Ni (1:20–40), co As
N	Gallium	Ga	15	Electronics	by Al, Zn
N	Cesium	Cs	7	Photomultipliers, space	co Li, Be
N	Germanium	Ge	7	Electronics, infra-red optics	by Zn
N	Beryllium	Be	6	Special alloys, ceramics, nuclear weapons and reactors, electrics	co Li, feldspar, Cb, Ta, Sn, mica
N	Scandium	Sc	5	Research (no known use)	by W, U, phosphate rocks
N	Arsenic	As	5	Biocides, enamels, glass	by Cu, Pb, Zn
N	Hafnium	Hf	4.5	Nuclear reactors, special alloys	by Zr (1:50)
N	Thallium	Tl	2	Electronics, biocides	by Zn, etc
N	Cadmium	Cd	0.5	Plating, pigments, batteries, plastics	by Zn
N	Bismuth	Bi	0.1	Drugs, cosmetics, alloys	by Pb, Cu, Zn, etc
N	Indium	In	0.1	Electronics, special alloys	by Zn
N	Selenium	Se	0.09	Electronics, glass, photocopying machines, pigments	by Cu (also Pb, Fe)
N	Tellurium	Te	0.002	Steels, rubber, chemicals	by Cu (also Pb)
F	Rhenium	Re	0.001	Special alloys	by Mo from Cu
M	Radium	Ra	0.0000013	Radiotherapy	by U (1 : 3,000,000)

Only a very few metals, such as gold and silver, ever occur naturally in their *free* or *native* state; most metals react with other elements, such as the oxygen in the air, to form chemical compounds. Those compounds that have a geological identity of their own, with a particular fixed arrangement of atoms, are called *minerals* and are the building-blocks from which rocks – and the earth – are made.

Some minerals are simple and made mainly of one metal; others are very complex and contain several metals together. Many of the thousands of known minerals are of academic rather than commercial interest: beryllium metal, for example, is an essential part of forty minerals and is found in fifty more[1], but of all these only one, beryl, has been found in mineable deposits. A few metals, such as rubidium – the ninth commonest metal in the earth's crust – are not a main constituent of any known mineral[1], occur only as impurities, and are thus produced only as by-products.

The more common minerals include oxides, sulphides, sulphates, carbonates, silicates, and the like: that is, compounds of a metal with oxygen, sulphur, sulphur and oxygen, carbon and oxygen, or silicon and oxygen. Water is also bound into the molecular structure of *hydrated* minerals. Chemical differences between minerals often make it possible to separate metal-bearing minerals from each other for smelting, the process that frees the relatively pure metal from the mineral.

Certain minerals are formed in similar conditions and hence tend to occur together. Many minerals of lead, for example, are found with minerals of zinc and silver; hafnium occurs with zirconium, and cobalt (and sometimes chromium) with nickel. Some mineral families are very complex; a relatively common one includes minerals of copper, lead, zinc, silver, gold, and traces of some fifteen other metals, all found and mined together with some emerging as by- or co-products of others.

Metal-bearing minerals can be formed by a wide variety of processes, and may change chemical or physical form or be redistributed in other processes. They can result from so many kinds of geological history, indeed, that in theory they can occur almost anywhere.

But this is not to say that they *do* occur almost anywhere – quite the contrary. Most of the minerals stuck together in tiny grains to form the earth's crustal rocks[2] contain only a few common metals: these minerals are combinations of oxygen and silicon (respectively 46.6 per cent and 27.7 per cent of the earth's crust by weight) with aluminium (8.1 per cent), iron (5.1 per cent), calcium (3.6 per cent), sodium (2.8 per cent), potassium (2.6 per cent), magnesium (2.1 per cent), titanium (0.4 per cent), and hydrogen, phosphorus, and manganese (0.1 per cent each). If you add all these up, you will find that all the other eighty elements together must make up less than one per cent of the earth's crust.

If all the atoms of any one of these eighty elements (mainly metals) were clumped together, we should have no trouble extracting them; but in fact they are widely scattered. Only rarely have geological processes, working over aeons, gathered into one place enough atoms of a particular metal to make a mineral concentration rich enough to be called an *ore* of that metal – that is, a deposit that can be mined at a profit. The formation of ore is an extremely rare geological event arising from unlikely accidents in the remote past. It is also, of course, such a slow process that though it is still going on, it cannot possibly replenish our supply of ores.

Ores vary greatly in *grade* – in the percentage of contained metal by weight[3]. Bauxite, the basic aluminium ore, is a mixture of three minerals, each made of aluminium, oxygen, and water in different proportions. Commercial bauxite is generally about half aluminium oxide, which is half metal by weight and half oxygen, and it is from this oxide – alumina – that the metal is smelted. The rest of the bauxite is impurities, mainly iron oxide and useless aluminium silicate. Thus the grade, the percentage of aluminium in recoverable form contained in the bauxite, is very roughly 25 per cent. In contrast, the world's largest silver mine[1] produces ore containing 165 g (about 5 ounces) of silver per ton of ore – a grade of 0.0165 per cent. This ore can be profitably mined only because the silver is a by-product of copper, lead, and zinc occurring at higher grades. Most big deposits of uranium ore grade at about 0.1 per cent; molybdenum ores, at 0.1 per cent; 'rich' concentrations of platinum-group metals in copper ores, at about 5 g per 1000 tons of ore (a grade of 0.0000005 per cent for all six platinum-group metals together, or of about 0.0000001 per cent for, say, iridium alone).

The distinction between the metal content of a mineral and that of an ore is important. Though the mercury content of cinnabar (mercuric sulphide, the usual mercury-bearing mineral) is 86 per cent, that of the cinnabar-bearing ores now being mined is only 0.2 to 0.5 per cent[1,4]. This is because small grains of cinnabar are scattered through large volumes of barren rock. And we are lucky to have that concentration to mine – the average concentration of mercury in the earth's crust (Table 1) is less than 0.00004 per cent.

Most ores[5,6] of the less common metals are thought to result from an upwelling of metal-rich magma (molten rock) or of metal-bearing hot water from deep within the earth's crust. Magma can solidify in place as metal-bearing igneous rock, perhaps after intruding into other kinds of rock, while the water can mineralize host rock (often in repeated waves of enrichment and intense fracturing) to form a *hydrothermal* deposit. Metallizing solutions often permeate cooling magma and surrounding rocks, perhaps forming both *massive* deposits – mineral veins along major fracture lines and along contacts (boundaries between different kinds of rock) – and *disseminated* deposits – mineral grains scattered along much smaller fractures within the rock.

These kinds of primary deposits can be modified in many ways. If the rocks are heated and compressed, metals can move about inside and become concentrated. Weathering can let water leach (dissolve) some metals, precipitate them elsewhere (secondary deposits), and leave others behind in more concentrated form (residual deposits). Physical erosion can carry rocks away as silt and sand, then redeposit the particles selectively (according to how heavy they are) in layers that can later harden into sedimentary rock. These and other processes – including biological concentration of metals, as by bacteria – can occur in various combinations and sequences, yielding very complex geologies: a single region might have manganese ores formed in three or four geologically different ways.

The general history of certain minerals may offer a clue to their physical form. Since bauxite is a residual ore formed by weathering, it is likely to form horizontal beds near the surface – unless later geological change has compressed and folded these beds, or changed them by heat and pressure into quite a different kind of rock (metamorphic rock). On the other hand, those low-grade (less than one per cent) copper deposits formed when metal-rich solutions mineralize plugs of crystalline magma that have intruded into other rocks and then solidified are likely to take roughly the form of a hollow cylinder, originally pointing (dipping) steeply downwards. These *porphyry* copper deposits – a term referring to fine-grained rocks containing coarser crystals – account for 90 per cent of US copper production, most of the world reserves, and a great many very deep openpit mines[1].

'Ore' is often loosely said to be any body of rock rich in metal-bearing minerals. But a particular body of such rock may or may not be of such a grade, size, position, and chemical composition that it can be mined at a profit – that is, called an 'ore' in the strict sense used in this book. Whether a body of mineralized rock amounts to an orebody[7] depends not only on its own characteristics but also on metal prices; on difficulties of transport and smelting; on local laws and conditions that may affect mining costs; on the price of labour and money; and on the wealth, political power, initiative, and ingenuity of the mining company. All these are bound to change in time; indeed, a deposit that is not quite an ore for one company may be made *payable* by another with a better idea. In the long run, as metal prices rise with increasing scarcity, deposits now too costly to mine will gradually become payable; hence any mineralized area, even a huge *mineral province* such as the Harlech Dome of West Wales, is potentially a mining area, and may become one sooner or later. Any submarginal deposit found there is just a nest egg with a long hatching time.

One physical, rather than economic, way of defining 'ore' is as a body of rock that bears metal at an abundance very much above its clarke. Composition may also matter: in present practice, aluminium oxide can be smelted (which is why bauxite is an ore) but aluminium silicate cannot (which is why, for the present aluminium industry, many rocks are just slag).

One might expect that as ore grades decline, available tonnages at those grades might rapidly increase. The classical form of such a distribution is said to have 'Lasky statistics', named after its discoverer. In fact, this agreeable distribution holds true only for a few metals found in widely distributed forms, such as aluminium, iron, and magnesium[4]. For some rarer metals, including copper in many porphyry deposits (though not in other kinds), it is also true over a limited range of grades – perhaps a factor (multiple) of ten or so[4]. But for most metals there is an abrupt gap, often of a factor of 1000 or 10,000, between the grade of the poorest mineralized deposit that exists and the clarke – that is, between the grade of the poorest deposit we can find and the grade of ordinary rock. Those metals for which very-low-grade deposits are probably small and widely scattered, if they exist at all, include (among others) mercury, lead, zinc, gold, silver, tungsten, tantalum, columbium, beryllium, antimony, and the rare-earth group[4]. For such metals, depletion is obviously serious, since smooth and continuous adjustments of price and technology are impossible if the distribution of grades is so discontinuous.

The distribution of many metals is spotty not only in grade but in location. The People's Republic of China, for example, controls about three-quarters of the world's known reserves of tungsten and one-half of the antimony; Spain, one-third of the mercury; the Republic of South Africa, three-quarters of the chromium and nearly one-half of the gold and platinum[1]. Since ores can be mined only where geological accident has put them, there is nothing that other countries can do about this.

Geological accident also makes some orebodies harder to mine than others. Those that stick up above the ground as outcrops are obviously easier to find and mine than those that are buried, *i.e.* blind or suboutcropping. For a given area and kind of overburden, the depth of the overburden determines the cost of stripping it away to expose the orebody. For iron ore today, the maximum economic stripping ratio (the ratio of the thickness of the overburden to that of the ore layer) is generally about 2:1, but some especially rich ores can pay for a ratio as high as 10:1[1]. Bauxite may even be mined at 15:1[1].

Geometry varies even more for ore than for overburden, and technique must vary with it. Oxide laterite ores of nickel, formed by weathering, are rarely more than 40 ft (13 m) deep, so they are simply stripped; horizontal US bauxite beds, similarly formed but thicker, can be openpit-mined to about 200 ft (66 m) deep; yet in France, the bauxite layers are no longer horizontal, but are folded into steeply dipping strata of limestone, so 80 per cent of the bauxite must be mined underground[1]. Near-vertical plugs of porphyry copper are often first exposed by stripping, then mined in an open pit, and finally – when the pit may be over 1000 ft (about 300 m) deep – pursued further underground. Such decisions are influenced by the ratio of poorly mineralized to richly mineralized rock (the *waste/ore ratio*), which is discussed in part 2 of Chapter Three.

The long, risky, and costly process of finding and evaluating orebodies (Chapter Three, part 1) is normally financed by risk capital left over from a mining company's past earnings. Exploration is a form of investment, and must, like any other, show a profit. No company will deliberately spend money discovering deposits that cannot be profitably worked within the company's time horizon for planning, which is generally two or three decades.

Once mineralization is found and studied, the mining company must decide whether and how mining could be made to pay. If the company wants to proceed and can get the necessary permissions, it must plan every detail of the mine in advance – nowadays an immensely complex, often computerized operation that requires intensive evaluation drilling and the skill and imagination of specialized consultants in many fields. Chapters Three to Five will describe the principles of mine design and the problems likely to arise both during and after mining; Chapters Six and Seven, the social, economic, and political implications of openpit metalmining.

But before going into these technical details, we must see how mining fits into a global pattern of resource depletion[7]. When we consume metals, we do not use them up by burning as we consume coal; instead, we change their distribution. Nuclear reactions and spacecraft excepted, the total amount of each kind of metal in the world never changes: what we do instead is extract metals from concentrated deposits, purify them, mix and fabricate them in manufacture, and either discard them as wastes or reclaim them as scrap. This constant flow of metals from one form and pattern to another is the subject of the next chapter.

Notes

1. See Introduction, note 3.
2. *Handbook of Chemistry and Physics*, Chemical Rubber Publishing Co. (Cleveland), 1961 (43d edn); data from Harrison Hale and from B. Mason: *Principles of Geochemistry*, Wiley, 1952.
3. The ores of a few metals, such as uranium, are sold and often graded by the weight percentage not of the metal but of the oxide or some other convenient compound.
4. See Introduction, note 4.
5. C. F. Park, Jnr and R. A. MacDiarmid: *Ore Deposits*, W. H. Freeman (San Francisco), 1970 (2nd edn).
6. C. A. Lamey: *Metallic and Industrial Mineral Deposits*, McGraw-Hill, 1966.
7. P. T. Flawn: *Mineral Resources*, Rand McNally (Chicago), 1966.

2

METALS AS RESOURCES

Because the earth is round, it is also finite, and the amount of each metal that exists on earth is strictly limited. Yet the price of each metal in a free market-place may have less to do with supply than with demand. Price depends on clarke only in an extremely broad sense – most explicit, of course, at the extremes of abundance and scarcity. At the high-clarke end of Table 1, prices tend to be low – in very approximate values[1] of £ per kg of metal, they are 0.30 for aluminium, 0.07 for iron, 0.40 for magnesium, and 1.40 for titanium. The differences reflect mainly the energy needed to unbind each metal from its minerals. At the other extreme, the low clarke of gold (£900/kg) and osmium (£5500/kg) inevitably makes them expensive.

Yet in between, where most metals lie, there is little relation between clarke and price[1]. Zinc is rarer than vanadium but about twenty-seven times cheaper. Lithium is four times commoner than lead but about sixty-five times dearer. The classic oddity is scandium, midway in clarke between molybdenum (£3.5/kg) and arsenic (£0.5/kg) – but costing about £6000/kg! This is because, despite much effort, no use for scandium has been discovered, so nobody is equipped to extract it; conversely, people cannot afford it. But such prices can drop: that of rhenium fell from \$10,000/g in 1928 to \$3/g in 1930, thus permitting its widespread use[2]. Price-swings the other way should, but do not always, decrease use: the price of mercury has risen 500 per cent in the past twenty years, and that of lead 300 per cent in thirty, but the use of both has greatly increased[1,3].

There is not even necessarily any clear relationship between price and demand for a particular metal over time[3]. Despite the efforts of trade groups to stabilize prices for tin and other metals, many prices still fluctuate violently for no obvious reason. This is especially true of secondary metals (Table 2). No matter how much demand there is for cadmium, if the only way to meet the demand is by producing more zinc (of which it is a by-product) than anyone wants to buy, the result will be a flooding of the zinc market and a severe fall in the zinc price, thus perhaps putting some zinc, and cadmium, producers out of business. For this reason, the cadmium supply depends on the output, and hence on the price, of zinc, not on the demand for cadmium. Likewise, prices of rare-earth elements fluctuate wildly because they are all co-products, but there is a shortage of some and a surplus of others.

Though we cannot deduce a metal's price from its scarcity, we can find deposits of it that are cheaper or costlier to mine than others. Indeed, economists often say that we can have as much as we like of the earth's stock of a given metal if we pay for it – true enough if we remember that the price may be completely out of the question[4]. Mining companies generally try to mine cheap ores before expensive ores – if they fail in this they cannot compete – and the idea that deposits will cost more to mine as they become leaner, deeper, and more remotely situated gives us a clue to understanding estimates of unmined reserves[3].

'Reserve' is an economic rather than a geophysical concept, and assumes today's technology and approximately today's prices. For a few metals, such as uranium, reserves can be estimated not only at today's prices but at various higher prices too; but the result is imprecise, since the cost of mining very-low-grade deposits several decades hence is anticipated only by analogy and extrapolation from known mining costs. Reserve estimates are thus inexact on both economic and geological grounds – the former because future mining costs are conjectural and the latter because companies do not deliberately spend money proving deposits that would be far too costly to mine today. But modern techniques of geological estimation are becoming increasingly sophisticated: 'indicated' or 'reasonably assured' reserves known to exist but not yet studied in detail and 'inferred' or 'estimated additional' reserves extrapolated to unexplored areas from experience in other areas geologically similar can often be estimated within a fixed range of possible error. Such techniques, averaged over whole continents or over the globe, eliminate most surprises. Many reserve estimates are precise to within the amount consumed in a few decades.

The US Bureau of Mines regularly projects[1] not only likely supply but also likely demand for each of eighty-eight minerals over the next thirty years. These global projections are not predictions, but calculations based on a series of alternative assumptions. The

USBM approach is to build scenarios for various specified 'alternative futures' on some stated general assumptions, including continued growth in population, GNP, and industrial production. All these alternatives together yield projected maximum and minimum values for future demand. Whether these levels of demand will actually occur depends, of course, on political and technological decisions taken meanwhile, but they are the best available guide to where demand could lead.

Demand projections may be compared with estimated world reserves to give some picture of how long demand at anything like present prices can be met with present technology. Two forms of comparison often used are the *static index* S – how long the known reserves will last at the present rate of consumption – and the *exponential index* E – how long the same reserves will last if demand grows at a rate G (in percentage/year compound). Table 3 gives global values[3,5] of these indices in years for sixteen major metals, as of 1970, with G the average of the USBM 'high' and 'low' projections to the year 2000. These values of G are often lower than present or recent values.

The unhappy reserve position shown in Table 3, though suggestive, is certainly not the whole story. The indices do not take account of new ore discoveries or of technological progress (both described below). Furthermore, mining companies make more effort to prove new reserves of some metals, such as mercury, than of others, such as gold; it does not always pay to spend a lot of money proving new deposits while you are still working old ones that seem to go on and on.

TABLE 3

KNOWN WORLD RESERVES OF MAJOR METALS

Metal	*G*	*S*	*E*	*Metal*	*G*	*S*	*E*
Aluminium*	6.4	100	31	Mercury	2.6	13	11
Chromium	2.6	420	95	Molybdenum	4.5	79	34
Cobalt	1.5	110	60	Nickel	3.4	150	53
Copper	4.6	36	21	Platinum group	3.8	130	47
Gold	4.1	11	9	Silver	2.7	16	13
Iron	1.8	240	93	Tin	1.1	17	15
Lead	2.0	26	21	Tungsten	2.5	40	28
Manganese	2.9	97	46	Zinc	2.9	23	18

* in bauxite

But the indices in Table 3 do correctly imply that if demand continues to rise approximately as projected, or even more slowly, most of the metals listed 'will be extremely costly 100 years from now'[5]. More detailed projections suggest that 'extremely costly' may mean costing between ten and a few hundred times as much as now.

This scarcity is all the more likely to occur because orebodies are so unevenly distributed that political limits can arise before physical or economic ones. Semi-monopolies such as the Chinese control of tungsten will become increasingly important and resource bargaining more bitter: wars over resources are not unknown. Some federations of producing countries could be very powerful: in 1970, for example, CIPEC (the copper-producers Chile, Peru, Zambia, and Zaire) produced 31 per cent of the world's copper[6] and 60 per cent of all copper exports, and held some 44 per cent of the known world reserves[1]. Recent strengthening of the equivalent OPEC oil group and the Chilean nationalization of US copper-mining give a foretaste of the next few decades: producing countries are starting to have second thoughts about turning their shares of the earth into someone else's dividends.

Monopoly need not be geographic or political. Some industrial cartels control large sectors of metal-markets, as with mercury. At particular times and places, subsidies or price controls may also distort market forces.

If we optimistically ignore political problems, then the depletion of the physical stock of any primary metal will follow the pattern studied by Hubbert for fossil fuels[3] and by Behrens for non-renewable resources in general[7]: consumption rises, peaks, and falls while prices tend to fall, stabilize, and rise steeply. Improved exploration and mining technology can delay the eventual skyrocketing of price, but the longer they do so, the faster the rise in the end; and after that, with reserves essentially exhausted down to very low grades, any residual demand will only be met at astronomical cost.

Before examining the possible escapes[3] from this pattern, one should note that many governments, seeking short-term gains, are hastening rather than delaying depletion. Subsidies for mineral exploration and extraction can include:

- outright grants (the UK Mineral Exploration Act 1972 offers mining companies £50 million as 35 per cent discretionary loans that are written off whenever a mine does not result);
- hidden subsidies for land, services, or equipment;
- 'tax holidays' (some countries impose a tax ceiling, set at zero for the first five or ten years);

- extra-low royalties (sometimes about one per cent or even less);
- 'depletion allowances' (the USA gives mining companies a staggering across-the-board tax write-off of up to 23 per cent to 'compensate' them, though not society, for running down their mineral deposits).

These concessions may not be intended to speed depletion, but this is often their effect, for they create a distorted market with artificially low costs and high demand. They also directly affect mining practice. A company whose earnings are exempt from tax for a set period has an incentive to compress its earnings into that period. Likewise, the pressure of the money market accentuates the need for quick returns on invested capital. A mining company may therefore prefer large short-term profit, so that it can maintain its fiscal strength and attract new investors, to smaller but longer-term profit. This may mean, for example, that the rich core of an orebody is ripped out as quickly as possible from a small, steep-sided pit – 'picking the eyes out' – though this makes large lower-grade deposits all round practically unrecoverable.

Few countries have a tax or market structure that encourages the full use rather than the hasty use of resources. (The South African gold-mining lease formula is one such attempt.) Probably no nation today recognizes mining's social utility enough to discourage mining companies from putting themselves out of business as fast as they can. This point is taken up in Chapter Seven.

As noted above, the reserves shown in Table 3 are not the same as resources or resource base – the total stock physically present in the earth's crust. Reserves can be enlarged, or demand for virgin metal reduced, by four panaceas that can delay the depletion implied in Table 3:

discovering new reserves of ore;
mining deeper or poorer deposits with improved technology;
substituting plentiful metals, or other materials such as wood or plastic, for scarce metals;
recycling (collecting and re-using) a larger proportion of the metal already being used.

All four of these panaceas, which Lovering examines in detail in his masterly survey[3], are encouraged by the rising prices brought on by increasing scarcity – the price mechanism well-known to economists. Not so well-known are its limits, sketched in the next ten paragraphs[3].

Many new orebodies, some of them major, will be discovered over the next few decades, some in traditional mining areas and most of the rest in remote or awkward places. Since most of the earth's land surface has now had at least a first reconnaissance, many or most deposits found from now on will be blind, not outcropping. Ever more elaborate prospecting instruments and more costly evaluation drilling will reveal deeper deposits which can be mined only at steeply rising cost in energy and land. These deposits will tend increasingly to be large, low-grade, and mineable only in open pits. They will also cause major land-use conflicts by occurring mainly in fragile, fertile, beautiful, or otherwise valuable places. We can't have our country, or our planet, and eat it too.

Under the sea-floor does not seem a promising place to search for ores[3]. Seabed nodules offer a very restricted cornucopia[3]. Nor do very deep mines on land offer a satisfactory escape; though the earth's crust is far deeper than we have yet penetrated (if it were a postage-stamp stuck on to a soccer ball, we would scarcely have got through the ink yet), nevertheless there are fundamental reasons to doubt the practicality of mining more than a few km down.

As resources are gradually depleted, the rate of new discoveries will fall despite rapidly increasing investment and better exploratory technology. For many metals this is already happening. Furthermore, the contribution of new discoveries to reserves will be blunted by the mathematics of growth at compound interest: if consumption of a resource is doubling, say, every seven years, as with aluminium, then new discoveries doubling the known world reserves will delay depletion by seven years. However, metals such as aluminium and iron, which are so overwhelmingly abundant that we shall always be able to mine them from silicates at a high, though not absurdly high, cost in energy and land, must be distinguished from rarer metals for which large low-grade deposits do not exist (see page 10). These more numerous non-structural metals are the ones to worry about.

Mining technology will continue – at ever higher cost – to improve. There might even be breakthroughs in, for example, methods of milling (crushing and grinding) rock to free the mineral grains. Yet no technology is going to make immense volumes of waste rock disappear, so the social returns of mining low-grade ores will continue to diminish. Mining technology will lose momentum sooner or later; it kept pace with declining grades of copper ores for many years, but has not been able to lower unit costs further since about 1958[3], and now they are probably rising. The unit costs of mining lead and zinc have taken an upward trend for some years[3] because technology has not improved as fast as grades have declined.

The direct energy requirements of mining and milling – already in the neighbourhood of 20 kilowatt-hours (electric) per ton of hard ore – will rise rapidly as more ore must be mined and milled for each ton of metal recovered. The steep rise has already begun[3]. Even if these energy needs can be met indefinitely, a point on which there is much doubt[8], it appears to be theoretically impossible, without greatly altering the earth's climate, to create the sort of energy-intensive cornucopian world in which metals are mined mainly from ordinary rock. The argument[9] leading to this conclusion is fundamental, based on physical law, and independent of new technologies.

Some metals can be and are substituted for others. Such metals as mercury and silver are being trimmed to essential uses. But substitution merely transfers pressure to a less-scarce resource, which is thereby made scarcer. Substitutes are not only likely to be inferior, but are also limited in number and ultimately in availability.

Two particular problems with substitution are better

appreciated by those on the cutting edge of technology than by some economists. First, some uses are non-substitutable. A really sensitive photographic emulsion of high resolving power can be made only with silver; no other atomic structure will do. Mercury, tungsten, the platinum and rare-earth metals, hafnium, zirconium, uranium, cobalt, nickel, antimony, beryllium, tantalum – all these and more have some unique functions that our immensely interlinked technologies could hardly do without.

Second, we know that it takes perhaps a decade to develop and 'debug' a new sub-family of alloys such as a series of stainless steels. It is not clear that we could repeat this task for large sections of metallurgy in a couple of decades, lacking some crucial metal. Even if a real economic scarcity in one metal might be tolerable, scarcity in several at once could be impossible to cope with in time.

Finally, recycling can be greatly expanded as a labour-intensive enterprise. Half our mercury is still lost as a highly toxic pollutant[10] rather than reclaimed; if we valued it as highly as we do gold and platinum, we could recycle almost all of it. But with most metals we do far less well, even where most of the available scrap is in an easily recycled bulk form: scrap brass is a hundredfold-richer copper ore than the ores now being mined, so for a given copper yield one needs only a hundredth the quantity. Urgent and close attention to recycling could take pressure off both mining and solid waste disposal: we are rapidly realizing that we have neither metals to throw away nor anywhere to throw them.

Large-scale recycling, now in its infancy, could realize more of its promise if subsidized by levies on the use of virgin metal. But recycling is not a long-term solution, for it does not create new reserves. It merely stretches the reserves we have, thus buying a little time in which to stabilize and reduce demand. And though effective recycling will usually save physical energy, it will demand the social energy to devise new market incentives and controls to ensure that when manufactured products have ended their carefully prolonged useful lives and must be scrapped, they are so designed that the constituent metals are in a recoverable form. Such new incentives must come sooner or later; the sooner they come, the less drastic and disruptive they need be. They are politically, technically, and economically more practical now than they will be if we wait until all other options have vanished.

In practice we shall apply, and have started to apply, all four of these panaceas in some combination. The result will be complex, but a recent computer simulation[7] suggests roughly what we might expect: new ore discoveries doubling the world reserves of any particular metal would typically delay the drastic spiralling of prices for a decade or two; both this panacea and the extensive mining of poor ores would slightly prolong the period of stable low prices, but would also shorten to a decade or two the period of transition to real scarcity; all four panaceas together, on a lavish scale, might support continued growth in, say, copper demand not for the twenty-one years shown in Table 3, but perhaps for as long as fifty or sixty years, leaving few options open thereafter.

Perhaps this is where we are now going, but it is surely not where we want to arrive. Our market incentives are making us grow now and pay later. Clearly, the only way to ensure continued supplies of essential metals for centuries hence, let alone for millenia, is to stabilize and then – if possible – to reduce both the number of people and the amount of metal used by each person[5]. We could then distribute the metals we have more equitably amongst all people[11,12] – there is simply not a big enough resource base to give everyone a Western European material standard. This brings us to the fascinating and controversial area of steady-state economics[13–16], to which unfortunately we cannot do justice here. Of course, we could still have growth in services, even if we stabilized the 'resource-intensive' sector of the economy: societies could continue to evolve – towards more sustainable goals and life-styles.

Limited resources, then, like limited ecological and social stability, imply that rich countries should do more with less, prefer simplicity and durability to elaboration and waste, maximize stocks rather than flows[14,15], and strive for stability rather than for gross affluence. Since lost time cannot be recycled, and since societies are slow to change direction, such reforms must be started decades in advance if they are to work at all. The alternative is stark[3]:

> When the time comes for living in a society dependent on scrap for high-grade metal and on common rocks for commercial ore, the affluent society will be much overworked to maintain a standard of living equal to that of a century ago. Only our best efforts in all phases of resource management and population control can defer that day.

Strictly speaking, even a thrifty society stable in total numbers and wealth, which is far from saying a stagnant or unrewarding society, will still have a small residual need for minerals. This is because recycling is never 100 per cent efficient, and the loss must be made up. But if demand can be stabilized well before reserves are exhausted, there is reasonable hope of meeting it for very many centuries. Table 3 shows that it is not so much the present level of consumption as the rate of growth in consumption that drains resources so quickly. The speed with which growth is destroying our options is what makes it so urgent that we adopt a prudent long-term resource policy based on conservation and efficient use.

A look at modern metal-mining in historical perspective, uncluttered by fantasies of nuclear utopia[8,9,17,18], reveals it as potentially as brief and transient as a flash in the pan. As Harrison Brown has pointed out, our elaborate metal-based industrial economy is unique and will not recur, for if it collapses, the highly technical means of extracting the only metal ores we shall have left will collapse with it[19]. Native copper, nugget gold, free tin, and meteoric nickel are no longer lying about as they were in the Bronze Age and before.

Thus those who value metal and deplore metalmania can take wry comfort from the thought that we cannot make the same mistake again. Yet even they will surely wish man to prolong the present experiment, and not to cut it short by a few generations' greed to consume everything in their own brief span[20]. It is from this imperative and this sobering perspective[3] that the modern trend towards meeting, and hence fuelling, demand for metals through large-scale openpit mining must be judged.

Notes

1. See Introduction, note 3.
2. See Chapter 1, note 2.
3. See Introduction, note 4, and Chapter 1, note 7.
4. The 10 billion tons of gold dissolved in seawater are extractable at a price – one so high that the very competent people and companies that have tried for decades to extract the gold have reportedly (Introduction, note 4) obtained so far a total of not quite 0.0001 g, worth less than £0.0001.
5. D. H. Meadows *et al.*: *The Limits to Growth*, Earth Island (London), 1972 and Universe Books (New York), 1972.
6. See Introduction, note 1.
7. W. W. Behrens III: The Dynamics of Natural Resource Utilization; R. F. Naill: The Discovery Life Cycle of a Finite Resource: A Case of Study of US Natural Gas; W. W. Behrens III and D. L. Meadows: Determinants of Long-Term Resource Availability; all in *Towards Global Equilibrium: Collected Papers* (D. L. & D. H. Meadows, eds), Wright-Allen Press (Cambridge, Mass.), 1973.
8. A. B. Lovins and W. C. Patterson: *Energy and Survival*, Earth Island (London), 1974 (in preparation).
9. A. B. Lovins: Thermal Limits to World Energy-Use, *Bull. Atom. Scient.*, Sept. 1973 (in press).
10. A. Tucker: *The Toxic Metals*, Earth Island (London), 1972.

11. Friends of the Earth: *The Stockholm Conference: Only One Earth*, Earth Island (London), 1972.
12. P. R. and A. H. Ehrlich: *Population/Resources/Environment*, W. H. Freeman (San Francisco), 1972 (2nd edn).
13. H. E. Daly: The Stationary-State Economy, *Ecologist*, July 1972, p. 4.
14. H. E. Daly: Toward a Stationary-State Economy, *Patient Earth* (J. Harte and R. H. Socolow, eds), Holt, Rinehart and Winston, 1971.
15. K. E. Boulding: The Economics of the Coming Spaceship Earth, *The Environmental Handbook*, Pan/Ballantine/Friends of the Earth, 1971.
16. H. V. Hodson: *The Diseconomics of Growth*, Earth Island (London), 1972.
17. H. Alfvén: Energy and Environment, *Bull. Atom. Scient.* 28, 5, 5 (May 1972).
18. A. B. Lovins: Testimony on Fast Breeder Reactors to the Joint Committee on Atomic Energy, US Congress, prepared for Friends of the Earth Inc.; excerpted *Bull. Atom. Scient.* 29, 3, 29 (March 1973).
19. Some economists who find a recent projection[5] of collapse unpalatable say that it ignores the price mechanism. This is incorrect: the results of the Behrens resource submodel[7] appear implicitly in the Meadows models[5]. Even more generous enthusiasm resides in those economists 'who, proceeding from analyses of sea-water or average granite, triumphantly calculate the enormous quantities of various elements contained in a cubic mile of sea-water or granite. This is a wholly frivolous and misleading approach which contributes only a false sense of

well-being and cannot withstand a thoughtful probe of the problem.' (Flawn: see Chapter 1, note 7.) Absolute resource limits matter only in a static and inanimate world. But the world is dynamic[5], with rates of change rapid compared to built-in delays. Such a system is inherently unstable. The instability can be reduced by slowing the rates of change, by reducing (via autocracy) those delays that arise from social rather than natural processes, or by the accurate long-range predictive planning that system dynamicists hope eventually to develop[7].

20. Herman Daly, discussing the morality of poor stewardship, points out that 'a physical steady state . . . absolutely requires moral growth. Future progress . . . must be made in terms of the things that really count, rather than the things that are merely countable.' Daly proposes marketable resource-depletion quotas auctioned by the government. Quotas will encourage thrifty technologies by making prices rise 'a bit ahead of schedule and more gradually. This . . . requires the [resource] optimist to live up to his faith in technology.' (A Model for a Steady-State Economy, Symposium on Limits to Growth, Yale University, Oct. 1972.)

3

BEFORE MINING

Part One: Exploration and Evaluation

Before mineral deposits can be mined, they must be found and evaluated. If the information so gained shows that a deposit can be mined at a profit, extensive data must be assembled and a detailed mining plan developed before any mining can begin. These interrelated processes – exploration, evaluation, and design – are the subject of this chapter.

Mineral exploration consists of studying the geological merits of smaller and smaller areas by more and more intensive methods. Rough surveys first eliminate unpromising areas; more costly, difficult, and exact methods are then concentrated on the rest. At each stage of exploration, both the cost and the likelihood of success increase as more and more prospects are dropped from the list and as those retained survive increasingly close scrutiny.

Exploration taken as a whole is mathematically analogous to gambling[1] at steadily worsening odds, so method and persistence can substitute for mere chance. Large investment skilfully applied has produced fairly consistent finds. Realizing this, most mining companies spend about a tenth of their gross income on exploration (which may be tax-deductible). Small companies must spend a larger fraction of their income, and a minimum of about £1 million a year, in order to compete[2], so they often form exploration syndicates. In recent years, the fraction that both large and small companies have devoted to exploration has risen as deposits have become harder to find[3]. Of course, money spent on research may substitute: metallurgical research can make payable many known deposits which would otherwise be useless by reason of their chemical composition.

Modern exploration does not depend on wandering prospectors who, with a 'nose for pay dirt', might stumble across something: it depends on careful section-by-section study of large areas, chosen either by geological analogy to known ore-bearing areas or by deduction from theories of ore genesis. Orebodies tend to occur in clusters, and the most promising areas are always those in which ores have been found in the past – particularly old mining areas.

Initial surveys often examine a block of up to 500 km^2 (about 200 square miles) for geophysical anomalies – places where instruments, either in light aircraft or on the ground, reveal a 'kink' in the earth's magnetic or gravitational field. Most anomalies turn out to be false alarms, but some may be mineralized areas. Another type of oddity often sought instead, or as a cross-check, is a geochemical anomaly – an abnormally high concentration of some metal or metals in local rocks and soil. Some mineralized deposits can be discovered and roughly located by mapping the metal content of sediment from streams into which metal or mineral grains have washed. Two other kinds of sensitive detective-work often used on the ground are botanical prospecting – some kinds of plants indicate or concentrate certain metals – and seeking the radioactivity of radioactive minerals or of their decay products.

Suspected deposits – for example, a place where a geophysical and a geochemical anomaly coincide – can be studied further by measuring the way in which an electric current, from batteries or a small dynamo, passes through the ground (resistivity and induced polarization studies). These techniques usually take a day or so and entail little more digging than would the planting of a few bulbs. In some cases, the vibrations from a small charge of high explosive set off in the ground are recorded with a network of microphones (seismic studies). This technique shows the thickness and stiffness of different layers of rock and soil, but it is rather imprecise. Shallow seismic studies are a moderate but short-lived nuisance to neighbours: the explosive charge is normally less than 1 kg – a couple of pounds – and a mechanical 'thumper' can sometimes be used instead.

All these methods provide only indirect, inferential evidence of what is under the surface. If, on this evidence, a prospect is still interesting, the next and more costly step is to extract deep rock samples.

This is normally done by rotary drilling with a special hollow drill-bit, the diamond-studded edge of which cuts in a ring, leaving a cylindrical core of rock inside. This core is removed in sections, laid in sequence in long cases, and analysed to find the composition of the various layers of rock. The drilling alone – not analysis – costs £2–3/ft (£6½–10/m). Sometimes, especially in soft rock, other drilling methods may be used, but with similar effect. About one or two per cent of the prospects on which drilling is begun eventually become mines[2], though if drilling goes on beyond ten or twenty holes it usually means that the first results are promising and that the odds of proving a payable orebody have greatly improved – enough, perhaps, to warrant spending a few hundred thousand pounds sterling on further study.

Like prospecting, exploratory drilling tries to narrow the area of interest. The extent of a deposit may be defined by drilling at increasing distances – perhaps in an X-shape – from the apparent centre of an anomaly. The mineralized area itself may be drilled on a grid pattern with spacing of perhaps 300 to 400 m (about 1000 to 1300 ft); such typical *scout drilling* has a fine enough mesh to intersect even fairly small orebodies, say 500 m across. In addition to the grid, or instead, a lone borehole may be sunk outside any systematic pattern. As scout drilling gradually reveals which areas may be mineralized and which probably are not, it may shade into a physically identical kind of drilling – prospect drilling – more specifically aimed at one small area, typically probing an identified deposit on a 100 to 150 m (about 330 to 550 ft) grid. Scout and prospect drilling are in practice often indistinguishable.

As the question asked about a successful prospect becomes not so much 'Is there an orebody here?' but more 'How rich and how big and what shape is it?', drilling may gradually enter the evaluation phase, aimed specifically at getting the detailed information needed to design a mine. Intensive evaluation may require vertical, or sometimes diagonal, drilling on a 50 m (about 150 ft) grid – 25 m in doubtful places – plus the sinking of a few prospect shafts several metres (say 6 to 10 ft) across in order to extract bulk samples for checking densities and for making pilot-plant studies of how the ore will be treated. Intensive evaluation is thus a major operation in its own right, the most costly and least risky phase of developing a mine. It typically takes a year or two, costs £1 to 2 million, is undertaken only after the company has decided in principle to open a mine, and includes 100 to 200 diamond-drilled holes plus some large boreholes and small shafts.

Most of the evaluation work is usually postponed until any necessary permissions for mining have been obtained, at least in outline, since otherwise the investment in the work could be wasted. But a company may get well into the evaluation phase, that is gather enough information to go far in its feasibility studies, without formally saying that it is doing any more than scout and prospect drilling. A company may in practice be able to begin basic pit design on the basis of several dozen boreholes; and with a hundred or so, nearly all the basic information needed for feasibility studies can be obtained. Even the proving of the huge Panguna orebody at Bougainville – one of the most complete evaluation programmes ever done – entailed only a little over 200 boreholes, totalling 274,000 ft or 84,000 m.

Exploratory diamond-drilling is usually done by a small diesel-powered drill rig, mounted on a 10 m (30 ft) tripod or on the back of a special lorry and run by a subcontractor's three-man crew. For a typical core drill-bit a few inches in diameter (say 4 to 8 cm), the diesel engine must be about 30 to 90 hp, depending on the depth and hardness of the rock. To keep the bit cool and to remove the rock particles or *cuttings*, water must be drawn from a stream or pond by a 10 to 20 hp diesel pump, pumped down the hole and back, passed through a tank so that the cuttings can settle, and then discharged. It may contain small amounts of oil or other chemicals. The maximum total area required for a mobile or semi-mobile rig, for storing casings (*i.e.* lengths of special pipe used to line the hole), and perhaps for a small shed is typically about 2000 m^2 (say $\frac{1}{2}$ acre). Most scout drilling goes down 100 to 500 m (about 300 to 1600 ft), corresponding to a typical time of between one and six weeks – in very hard rock, perhaps fifteen or more. Drilling to these depths does not normally need a concrete *pad* or working foundation such as is needed for deeper drilling or larger holes, as for oil and gas.

With properly enforced precautions against oil spills, water pollution, litter, fumes, and damage to ground-cover, drilling need not do substantial permanent damage to land. The capped borehole itself is usually hard to find after a year, though scars from hauling equipment may take much longer to heal. The main annoyance of careful drilling is likely instead to be noise, nearly continuous during working hours, which may be round the clock unless restricted; the main importance, as a possible prelude to a large mine (see part 4). That possible sequel, not the immediate nuisance of drilling, is the real issue.

Part Two: Design

Designing a modern openpit metal-mine is an extremely complex job that can take teams of experts several years. Yet the general principles of design are relatively simple, and this chapter will enable you to do your own (very approximate) design estimates from basic data. We shall start with the physical details of the mine – what they are and how they are designed – and then briefly discuss other matters that must also be planned – transport, marketing, finance, and permissions. Labour and the after-use of the land will be discussed in Chapters Six and Five respectively.

An openpit metal-mine comprises a large hole in the ground; equipment for drilling and blasting large blocks of rock inside the pit; equipment for gathering up the broken rock and hauling it to a pitside crusher and further stages of a mill that reduces the rock to powder; a concentrating plant that sorts out the desired mineral grains from the *gangue* (unmineralized rock), yielding a concentrate and a much larger mass of tailings or slimes; a tailings pond or lagoon, confined by a tailings dam, in which the slimes can be permanently impounded; a place where overburden and unmilled waste rock can be tipped; support facilities for the various people and machines; means of transporting supplies in and concentrate out; and sources of water, electricity, and fuels. We shall cover all these points later, but for the moment shall concentrate on the basic process – moving large masses of rock – and shall ignore, among other things, the operation of the ore-dressing plant (mill and concentrator). We shall merely assume this plant to be a very large black box with lumps of mineralized rock going in one end and with concentrates and tailings coming out the other.

The purpose of an evaluation programme is to tell the mine designer the grade, and indeed the exact chemical composition, of every part of the orebody. He can then draw up tables and graphs, build physical models, and even feed the data into a computer that can simulate the operation of various proposed mining schemes and calculate which will pay the best in the shortest period of time.

The reason a computer is useful in this feasibility calculation is that the grade of ore varies, often by a factor of two or more, between some parts of the orebody and others. A richer part must often be worked first, out of logical sequence, in order to recover capital costs quickly; or perhaps richer and poorer ore must be worked simultaneously and then blended to maintain a constant average grade going into the ore-dressing plant; or there may be other design constraints on the pattern of working. Though our rough estimates in this book will use only average grades for entire orebodies, the exact spatial distribution of grade throughout each orebody is important to the professional mining engineer.

The two basic characteristics of any orebody are total tonnage and average grade (chemical composition matters to the team of consultants designing the ore-dressing plant, but for the moment we are ignoring their problems). For example, a low-grade disseminated copper deposit might be said to contain 70 million tons – about the mass of 100 large aircraft carriers – grading 0.60 per cent Cu; another might contain 280 million tons grading 0.71 per cent, surrounded by a fuzzy 0.28 per cent deposit amounting to some 45 million tons. The former specification is the 1967 content of the Bethlehem reserves[4] in the Highland Valley of British Columbia, a cluster of small 'plums-in-pudding' orebodies near the minimum economic size; the latter specification might be typical of a very large mine similar to Palabora in the Transvaal. These grades are, I repeat, averages: the Bethlehem 0.60 per cent deposit contains both ore richer than 0.80 per cent and ore poorer than 0.40 per cent, and may go considerably outside both those values[5].

Two further characteristics of any orebody are crucial to the designer: the waste-to-ore ratio and the expected recovery of the concentrator. The waste/ore ratio shows how many tons of waste rock (rock which is below the cut-off grade and thus considered too poor to mill and concentrate) must be removed, on average, in extracting each ton of ore. Waste/ore ratio depends on the cut-off grade, the shape and working pattern of the pit, and the shape and structure of the orebody; and all but the last two of these characteristics can be varied. Recovery – what percentage of the metal contained in the ore going *into* the concentrator is contained in the concentrate coming *out* (the rest is lost in the tailings) – depends on the composition of the ore and on the skill and money applied to making and running the ore-dressing plant. Though we shall express waste/ore and recovery, like tonnages and grades, as rough averages, the designer must know these figures for every part of the orebody and for every month in the planned life of the mine.

Note that cut-off grade, reserves, and grade are not independent data. For example, an orebody with higher-grade zones merging into progressively poorer zones, as is common with disseminated deposits, might contain 100 million tons grading 0.50 per cent if we count everything down to 0.10 per cent; or only 25 million tons at a richer 1.0 per cent or more if we count only the richest part (0.50 per cent cut-off); or, more conventionally, 50 million tons at 0.80 per cent assuming 0.20 per cent cut-off. With luck and ingenuity, all of this reserve may be extractable; if it is the wrong shape, though, a small part of it might be unrecoverable without incurring a huge waste/ore.

Here is how our average figures can work: the Bethlehem ore[4] (Jersey pit) had an average waste/ore of two (twice as much waste as ore) during the first few years of working because most of the waste was on top; but after that was removed, the ratio dropped to a very low 0.25 or less, yielding an average of 0.655 for the whole pit[4] – an unusually favourable figure for an openpit copper-mine. (Most such mines have waste/ore of at least one; the US average[6] is 3.5; some have between four and five. If waste/ore exceeds five or so, underground block-caving mining is often cheaper[7].) At this ratio of 0.655 tons of waste per ton of ore, for the 70 million tons of ore mentioned earlier there must be 46 million tons of waste, or a total of 116 million tons of rock to be removed from the pit. All this rock-moving yields, with our 0.60 per cent grade, 0.0060 × 70 million tons, or 420,000 tons, of copper going into the concentrator. If we assume that 83 per cent of this copper is recovered in concentration – the 1970 US average recovery for copper in all kinds of ore was 83 per cent[6] – then the total copper output of the mine over its lifetime will be about 350,000 tons of copper in concentrate – a small openpit mine by modern standards.

Exact recovery figures for any particular mine are generally a trade secret and depend on local circumstances. But we can get a good rough figure by taking the 1970 averages for the entire US mining industry[6]. Table 4 gives these averages for each of nine families of metalliferous ores; it also shows the average grade of these ores and of the concentrate, or, for complex ores, of the various separate streams of concentrates, emerging from the ore-dressing plant. In general, the best modern recoveries might be 5 to 10 per cent higher than the averages shown. Concentrate grades may vary more widely, perhaps attaining a maximum of 40 per cent for copper[8].

If we apply the Table 4 data for plain copper ores to the Bethlehem mine – 80 per cent recovery into a concentrate grading 23.3 per cent copper – we can compute a total copper output of 336,000 tons contained in 1/0.233 times that much, or 1.44 million tons, of concentrate leaving the mine. The 20 per cent unrecovered copper, 0.20 × 420,000 tons = 84,000 tons of it, will be lost in the tailings. How much tailings? Well, just the original 70 million tons of ore minus the

TABLE 4

ORE GRADE, CONCENTRATE GRADE, AND CONCENTRATOR RECOVERY
(US average, 1970)

Ore or Concentrate	Grade of *Ore* or Concentrate Cu (%)	Au (ppm)	Ag (ppm)	Mo (%)	Pb (%)	Zn (%)	Fe (%)	Concentrator Recovery (%) Cu	Au	Ag	Mo	Pb	Zn	Fe
Cu	*0.87*	*0.089*	*4.5*											
Cu	23.3	2.08	114					80	71	83				
Cu–Mo	*0.68*	*0.18*	*1.9*	*0.020*										
Cu	25.3	4.7	79					84	64	76				
Mo				53.7							59			
Cu–Pb–Zn	*0.40*	*0.64*	*178*		*4.2*	*1.2*								
Cu	25.5	5.7	6350		6.2	6.0		77	40	87		1	6	
Pb	0.88	17.5	138		72.8	1.6		3	27	8		95	3	
Zn	1.06	0.16	302		1.8	54.4		1	5	2			70	
Cu–Zn–Fe–S	*0.82*					*1.89*	*33.3**							
Cu	18.6	0.33	39					90						
Zn						50.6							61	
Fe							91.2*							80
Au–Ag		*20.8*	*187*											
Au–Ag		1985	17520						94	92				
Pb–Zn					*3.65*	*3.40*								
Pb			44		74.8							97		
Zn			332			56.8							87	
Pb–Zn–Ag	*0.35*	*2.5*	*170*		*4.40*	*6.08*								
Pb	3.10	*20*	1623		57.5	7.41		70	79	83		87	9	
Zn	0.65	21.1	129		4.0	53.2		14	6	8		8	82	
Zn						*4.99*								
Zn						61.0							93	
Fe							*33.7*							
Fe							63.9							83

* expressed as % iron sulphide

TABLE 5

TYPICAL DENSITIES (TONS PER CUBIC METRE) OF VARIOUS ROCKS

Basalt	2.95	Copper ores	2.6–3.2	Limestone	2.2–2.6	Sand*	2.0
Bauxite	1.9	Earth*	2.0	Manganese ore	3.1	Sand & gravel*	2.25
Chalk	1.85	Granite	2.4	Phosphate rock	2.0	Sandstone	2.5–2.65
Clay*	1.7–2.1	Gravel*	2.1	Quartz	2.6	Shale	2.4
Clay & gravel	1.8	Iron ore 40%	2.65	Salt	2.1	Taconite	4.75

* wet; for dry values, subtract 0.3

1.44 million tons of concentrate, or about 68,560,000 tons! You can see how 'concentrate' gets its name.

You may now want to convert some of these tonnages into volumes, and Table 5 gives the necessary data – the typical density, or weight per unit volume, of various kinds of rock in place[9]. If you are not sure of the proper category for a given rock, assume a density of 2.7 ton/m^3 for hard rock and 2.3 for soft, and you will not be too far wrong. But rocks have been compacting for millions of years: if you break them apart, you cannot pack them together again in the same volume. The extra space they take is expressed by a *swell factor* ranging from about 1.2 for materials already full of airspaces or voids, such as sand or gravel, to 1.9 for dense, compact rocks such as basalt. Mine designers work from exact measurements, not estimates like these, but for us it is a good rule of thumb that 1 m^3 of solid rock makes about 1.6 to 1.9 m^3 of broken rock, which then contains about 40 to 45 per cent voids. A good average swell factor for hard rock is 1.75. When broken rock becomes tailings, it is a sort of fine sand that packs together better but is also roughly half water; so because of the extra volume of the water, the total swell factor from bank, or bulk rock, to tailings is about 2.2 to 2.4.

Now to put these figures to work. If the density of rock at the Bethlehem mine is, say, 2.5 ton/m^3 (a little denser than granite), one ton of waste rock must occupy 0.4 m^3 before it is broken and about 0.7 m^3 afterwards, assuming a swell factor of 1.75. Thus the 46 million tons of Bethlehem waste would occupy just over 30 million m^3. Of course, the more steps are in our computation, the more our errors might build up, but we are still unlikely to be wrong by as much as 40 or 50 per cent. Since we are not designing a mine, this margin of error is not too great for our purposes.

So far we have been using ore characteristics which the designer must take as he finds them: tonnage, grade, composition (which determines the practical limit of recovery), and waste/ore for a particular choice of cut-off grade and pit shape. But the designer can choose not only the pattern but also the rate of mining: the mine *life* and the concentrator *throughput* (the daily tonnage of ore being milled and concentrated, averaged over every year of the mine life – each year may have 300 to 360 working days). All these factors are interrelated; thus throughput must rise if life or cut-off grade falls. The designer, aided by his computer, must therefore adjust all the variables to see which values will minimize production cost and thus maximize profit. Of course, short-term profit may still mean long-term loss, and every design change also changes the cost of producing a ton of metal; so the planner must normally use discounted cash-flow analysis – a way of evaluating future costs and profits taking into account the cost of all the needed capital. Since the planner cannot know in advance how the metal market or the value of money will fluctuate, all he can do is see how sensitive profit will be to financial variables. This will give him some idea of the risk in the investment.

The most profitable lifetime for an openpit metal-mine with throughput of about 20,000 ton/day or less is normally about twelve to fifteen years. The life of a very large mine may be fifteen to twenty or even twenty-five years: seldom longer, since if production is stretched out too long, the capital would earn more by being banked at compound interest. For example, while the relatively small Bethlehem copper-mine (Jersey and East Jersey pits only[4]) is scheduled for about twelve years, Palabora and Bougainville will run for twenty-five or thirty, and may even continue underground after that. But they are among the largest copper-mines in the world (Table 6), and only in their size range is the optimal life likely to exceed fifteen to twenty years.

Cut-off grades are now a few tenths of a per cent in copper-mines, varying widely with other kinds of ore – 0.20 to 0.35 per cent are typical for copper, but some large mines are nearing 0.10 per cent already, and that may be the norm by 1980. Already, mines are being designed so that the wastes below cut-off can be recovered later and their copper values extracted, perhaps by chemical leaching. In other words, wastes are being stockpiled, not just dumped. Even the 0.12-per-cent-and-under wastes of the big US copper-mines

TABLE 6

TWO VERY LARGE OPENPIT COPPER-MINES*

Name	Palabora	Bougainville
Location	northeast Transvaal	Territory of Papua & Nev Guinea
First production	1964	1972
Reserves (million tons)	315	900
Grade	0.68% Cu	0.48% Cu + 0.55 ppm A
Working cut-off grade	0.25% (0.20% before 1971)	
Average waste/ore (approx.)	1.3	0.7 (unusually low)
Average throughput (t/day)	50,000–60,000	80,000
Average Cu recovery	84%	95% (unusually high)
Average concentrate grade	34% Cu	30% Cu + 34 ppm Au
Average annual production (tons)	100,000 Cu; 1,000,000 magnetite conc.; 80,000 sulphuric acid; + U, Th, Zr, etc	150,000 Cu; 15 Au; 31 Ag
Output shipped by	rail (smelted onsite)	5½-inch × 15-mile pip ship
Total waste (million tons)	700 (estimated)	400
Overburden (million m^3)		28 (average depth 15 m)
Haulage fleet (trucks)	24 58-ton, 10 90-ton	41 100-ton (1000-hp)
Final pit area (million m^2)	0.3 (estimated)	0.5
Final pit depth (m)	490	885
Underground continuation	at least 430 m	unknown
Total site area (million m^2)	20.5 leased	95 leased, plus unknow offshore area affecte by tailings
Total labour force (approx.)	3100 (includes smelting, semi-fabrication, by-product processing)	3200 (10,500 peak durin construction phase)
Total capital cost (approx.)	£65 million (unusually low)	£200 million (unusuall high due to remotene and steep terrain)
Typical pretax annual profit	£20 to £30 million	£30 to £50 million

* Sources: Palabora Mining Co Ltd, *Annual Report and Accounts 1971*; Palabo Mining Co brochure in English/Afrikaans, undated, late 1960s; Bougainville Minir Ltd, *Annual Report 1971*; *Bougainville Copper . . . an introduction*, undated, probab early 1970.

(thirty-five milling more than 1 million tons of ore a year and eight exceeding 10 million) may soon be a profitable source of copper.

Once the mine designer has decided the best pit shape, life, throughput, cut-off, and shift schedule, and has tabulated the tonnage and grade of ore to be worked at each level of the pit during each month of its life, he can decide how many of what kinds and sizes of equipment will give him the most work for his money. His choice of equipment – each combination with different capital cost, operator cost, and maintenance cost – will determine, and will be determined by, the pit design he has chosen, and the resulting costs may also influence his choice of life, throughput, and cut-off – a complex business! But what will it look like when built?

Ordinarily, an open pit in hard rock is rather oval or oblong in plan (orebodies are seldom perfectly round), starting as a ring-shaped trench and ending as a deep dish. The bare rock sides of the dish will slope downwards and inwards at an average of about 45°, depending on the shape and physical properties of the orebody; steeper walls might fall in. Instead of being smooth so that rocks could roll down, the sides will be divided into steps or *benches*, horizontal terraces perhaps 10 to 15 m (30 to 50 ft) high – 10 m is typical of small mines, 15 m of large ones. Every second or third bench might be cut extra wide to catch rockfall and for increased stability, especially if the heavily fractured and faulted rock common in mineralized zones is not sufficiently firm and stable. Each bench is separated from the next by a *face*, which is just like the riser in a flight of stairs but 10 to 20° less steep. A mature pit has concentric benches running round it and converging on the centre, and, viewed from directly above, might look like a giant thumbprint. The biggest such pit in the world – at Bingham Canyon, Utah – is about $1\frac{1}{3} \times 1\frac{2}{3}$ miles $\times \frac{1}{2}$ mile deep (2.1 × 2.7 × 0.8 km).

The pit is worked by drilling a row of holes about 6 to 7 m (20 ft) apart at a *working face* that is below, and hence further towards the pit centre than, the faces and benches already completed. This face, when completed from the working platform of the bench above it, will define the outer edge of a new bench below it. Cuttings extracted during the drilling are analysed to assist grade control in the ore-dressing plant. Once a day or once a shift, explosives packed into the row of holes blast to rubble a block of rock, typically 10 to 15 m high by 6 to 15 m deep by 50 to 300 m wide. The broken rock is then scooped up by electric shovels or similar machines[9] and hauled to the nearby crusher[10]. Because the location of the working face, of which there is often more than one at a time, changes constantly, fixed ore-moving devices such as conveyors, cableways, and buckets are seldom suitable ore-haulers, even if they can handle large pieces, so fleets of very large dump-trucks are normally used instead on a special network of carefully maintained hauling roads, often 15 to 25 m wide.

The designer must draw up a complete chart of material flows to and from each part of the mine during each month of its life, and must then compute how many and what size of equipment is needed to keep a steady stream of ore moving. As the pit deepens, the haulage fleet may have to increase; worn-out equipment must also be replaced. The capital cash-flow of an openpit mine is thus very complex. Innumerable facts of mining life limit the designer's freedom. For example, the pit must be large enough so that the maximum gradient of uphill hauling roads will not exceed about 8 per cent, lest laden ore trucks be unable to climb it. On such a slope they might do 6 to 8 miles per hour and be heard for miles.

The capital cost of the needed machinery is staggering. A single 6-yd^3 electric excavator costs about £180,000; a 6-inch (15-cm) drill for blasting, £40,000; an 85-ton ore-carrier some £90,000. Capital cost usually goes up almost – not quite – in proportion to size, yielding economies of scale; 25-yd^3 (19-m^3) shovels and 150-ton trucks are in regular use, with larger versions under development. Steady, reliable operation is essential to keep costs within bounds: at Palabora, every forty seconds, day and night, a 58- or 90-ton truck must dump a load of ore into the crusher. The cost of reliability is high. Ore trucks last only four to six years; tyres, a few thousand hours at £3000 to £8000 a set; a 6-inch drill capable of boring 9 to 12 m (30 to 40 ft) in an hour, five years; and the £70 drill-bit, only 200 m – about sixteen holes – in hard rock. Even the slurry explosive, typically ammonium nitrate and fuel oil ('AN-FO', often combined with a solid high explosive), is not cheap: it usually runs about 2p per ton of rock, compared with $\frac{3}{4}$p for the drilling. It takes about 300 to 600 g of explosive to break 1 m^3 of rock. And just to keep the mine running takes numerous support vehicles: crew buses, runabouts, bulldozers, small loaders, graders, cranes, and trucks for carrying tyres, grease, fuel, water, explosives, or repair crews.

The ore-dressing plant at such a mine is a voracious consumer of water, electricity, reagents (chemicals used in concentrating), and steel: Table 7 gives 1970 US average consumption figures for dressing various ores[6]. Such data vary widely from one mine to another, but average values still give a useful impression of magnitudes. For example, it appears that milling and float-concentrating one ton of a typical non-ferrous metal ore takes roughly 20 kilowatt-hours of electricity, 3000 litres (3 m^3 or 3 tons or 660 Imperial gallons) of water, many hundreds of g of steel, and roughly twice that quantity of special chemicals worth some £0.05 to £0.10. To understand why these figures are so large, we must now find out how the plant works.

It would cost too much to transport low-grade ores to a smelter; they are therefore concentrated at the mine, as close to the pit as possible, by a factor of about 32 for the average sulphide ore and about 2.3 for the average oxide or carbonate ore[6]. Oxides may have to be leached, for example with sulphuric acid, before or after they are concentrated. Various methods can be used to separate the desired mineral grains from gangue: magnetism for iron, strong electrostatic fields for some other metals, gravity for very heavy metals such as gold and tungsten. But the most common method is called selective flotation or froth flotation. The principle is to crush and grind the ore to powder; coat the grains with a special oil which, because of the chemical nature of the surface of the metal-bearing mineral grains, adheres only to them; and put the mixture in

TABLE 7

CONSUMPTION PER TON OF ORE CONCENTRATED BY FROTH FLOTATION PLANTS
(US average, 1970)

Ore	*Electricity (kW-hr)*	*Water (litres)*	*Steel (grams)*	*Reagents (grams)** *mod.*	*act.*	*dep.*	*col.*	*fro.*	*flo.*	*other*	*Reagent value (£)*
Cu	17.6	3130	724	4030	39	3	44	24	3	—	11.4
Cu–Mo	16.1	2980	1034	1590	—	27	32	41	14	1.5	8.4
Cu–Pb–Zn	18.6	2130	291	500	55	313	58	20	6	—	14.2
Cu–Zn–Fe–S	23.9	3360	470†	4550	358	96	155	48	2.5	—	21.7
Au–Ag	§	§	200†	1250	—	—	132	20	14	—	9.7
Pb–Zn	21.5	3025	289	442	207	134	76	44	3.5	30	22.9
Pb–Zn–Ag	29.3	3380	1318	1040	469	282	137	89	7	40	44.0
Zn	15.8	2090	243	1680	247	35	34	30	—	—	16.0
Fe	32.2	8180	1264	2190	—	172	489	12	8	—	9.5

* mod. = modifiers (lime, phosphates, sulphuric acid, sodium carbonate, sodium hydroxide, sodium silicate, sulphur dioxide, Calgon, soda ash, etc.)
act. = activators (sodium sulphide, copper sulphate, etc.)
dep. = depressants (dextrine, phosphorus pentasulphide, sodium cyanide, sodium dichromate, sodium hydroxide, zinc sulphate, starch, sodium sulphite, sodium fluoride, etc.)
col. = collectors (various xanthates, thiocarbamate, fuel oil, special reagents)
fro. = frothers (pine oil, cresylic acid, creosote, methyl isobutyl carbinol, special reagents)
flo. = flocculants (special reagents)
other = other reagents, often proprietary
† Estimates based on partial data
§ Data unavailable

a flotation cell – a large tank where, under carefully controlled chemical conditions, rising air bubbles make a layer of sticky foam which traps the oil-coated grains while the rest – the unwanted gangue – sink to the bottom. The froth, with the metal-bearing mineral stuck to it, then spills over the side of the cell into traps, and the mineral is filtered, washed, and dried. The gangue is thickened – the excess water decanted – and piped to the tailings pond; it may be scavenged first to remove all payable products, perhaps including by-products such as iron- and sulphur-bearing minerals. In a real concentrating plant, ore passes through many cells in succession, often returning for further grinding in between. The huge array of cells required – very roughly 15 to 25 litres (0.5 to 0.9 ft^3) of cell capacity for each ton/day of copper ore throughput, for example – may fill a building of several hundred thousand m^2 (several million ft^2); yet the entire ore-dressing plant, largely automated, can be run by a handful of men.

Efficient froth flotation is so difficult that the ore-dressing plant typically costs between a third and a half of the entire capital cost of the mine – so much that the plant must run all the time. This is partly because it is so difficult and energy-consuming to reduce rock to powder – the smaller the mineral grains, the finer the powder needed and the greater the energy needed to grind that fine. A primary crusher, such as a huge rotary hammer or nutcracker-like jaw, must first break large blocks into pieces no more than about a hand's-breadth across; then a battery, often conveyor-fed, of secondary crushers using specially hardened metal cones can break the rock to pieces perhaps a centimetre ($\frac{2}{5}$ inch) across; tertiary crushers of similar design may be needed too. Once the rock is in fairly uniform pieces like small grapes, it can be fed wet into powerful mills in which steel balls 2 to 3 inches (5 to 8 cm) in diameter pulverize the rock inside a steel tank or *liner.* Both the liner and the balls are gradually worn away. Rod mills may be used instead: they use less steel, approximately 400 g or $\frac{1}{2}$ pound per ton of ore (a third as much as for ball mills), though they also grind less finely. Mills are very active machines: a ball mill 18 ft in diameter by 21 ft long (6 × 8 m) can mill daily about 10,000 tons of typical copper ore – but the grinding action powered by its 4250-hp motor consumes some 5.5 to 7.5 tons of balls daily. (The steel consumptions listed in Table 7 are totals for balls, rods, and liners.) Some ores can be ground autogenously, using the ore itself as the grinding medium, but that usually takes much more energy.

Flotation cells also need power for stirring, but their main running cost is for water and reagents. Large cells – hundreds of ft^3 (the largest[6] now in use is about 2000 ft^3 or 57 m^3) – may use water, reagent, power, and money more efficiently than small cells, but they still need a lot. The chemical balance in the cell depends mainly on modifiers – chemicals that control alkalinity, precipitate suspended colloids such as fine clays, and bind soluble salts that would otherwise interfere. Lime, used in immense quantities with acid ores such as pyrites and other sulphides, is a typical modifier. Other reagents used in smaller but more critical quantities include oily collectors to help some mineral grains to stick to bubbles, frothers to make the bubbles, activators to improve the 'floatability' of certain desired minerals, depressants to make unwanted minerals sink faster, and flocculants to make the suspended solids clump together for easier thickening and filtering. Some of these reagents are extremely powerful – a good frother produces hundreds or thousands of times as much foam as a household detergent – and even small traces of them in the incoming water can upset the efficiency of flotation.

For this reason, water can be re-used only a few times before it has collected so much reagent that the cell cannot work properly. Used water must be constantly drained off and makeup water added. All year round, even in the driest summers, a modern metal-mine must have enough pure water, and even in rainy areas this usually requires a special reservoir. Careful design can economize on water; but the US mines represented in Table 7 are mainly in arid regions where this has already been done. (The water consumption shown is for re-used *plus* makeup water.) As a bare minimum, a concentrator with maximum water recycling still needs makeup water of at least one ton (1 m^3) per ton of copper ore processed, and similarly for other ores. Depending on the ore chemistry, twice this much makeup water might actually be needed; or, if little water can be recycled from tailings, twice as much again. Any reservoir must drain a large enough catchment to receive and store a correspondingly large amount.

Part Three: Design Examples

To fix ideas, let us pause here to give a few more numerical examples of how to estimate mine size. Estimate, not calculate or predict: as amateurs we cannot expect any but approximate results.

First, suppose we have a typical small openpit lead–zinc mine with recoverable reserves of 12 million tons grading 3.6 per cent Pb and 3.1 per cent Zn; this is similar to several mines now operating in the Irish Republic. Suppose further that waste/ore is a favourable 0.8 (if it were larger, the mine would be larger too). The waste rock to be moved is thus 0.8 × 12 million tons = 9.6 million tons. If the 12 million tons of ore can be concentrated with a recovery (see Table 4) of, say, 95 per cent for Pb and 85 per cent for Zn, either in two separate concentrates or in one combined concentrate, then we can recover a total of 12 million tons × 0.036 × 0.95 = 410,000 tons of Pb and 12 million tons × 0.031 × 0.85 = 315,000 tons of Zn (as usual, we are using round numbers). If we obtain a concentrate grade of, say 75 per cent Pb and 50 per cent Zn, the total amount of concentrate produced will be 1,180,000 tons – say 84,000 tons a year to be shipped for fourteen years. Twelve million tons of ore less 1.18 million tons of concentrate equals 10.82 million tons of tailings. The tailings also contain a negligible amount – say 10,000 to 20,000 tons – of 'consumed' steel and 'lost' reagents. Average annual production over fourteen years will thus include about 52,000 tons of metal in concentrates, 770,000 tons of tailings, and 690,000 tons of unprocessed waste rock.

To calculate the approximate areas of the pit, waste tips, and tailings pond, we need merely assume the reasonable average figures of 1.7 for swell factor from bank to loose rock and 2.3 from bank to tailings. The pit is the hole formed by the removal of 12 million tons of ore and 9.6 million of waste. In bulk form, these 21.6 million tons of rock, with a probable density of 2.7 ton/m^3 (a bit denser than limestone – most lead–zinc deposits occur in limestones or dolomites), would occupy 8.0 million m^3. If a pit of this volume were shoebox-shaped, with vertical sides, it might be, for example, 13 hectares in area (33 acres) and 60 m (200 ft) deep. With sloping sides – after all, they could not possibly be vertical in practice – the area would be larger; a typical computation is discussed below. The 9.6 million tons of waste, having swelled 1.7-fold, would fill some 6 million m^3. In a vertical-sided block 60 m high, it would occupy 10 hectares (25 acres), and with sloping sides, perhaps at least 50 per cent more. Finally, the 10.8 million tons of tailings would occupy 9.2 million m^3 (equals 10.8 million tons × 2.3 swell factor ÷ 2.7 ton/m^3). The tailings pond might thus be, for example, 46 hectares (114 acres) in area and filled to an average depth of 20 m (65 ft).

We can thus estimate the area needed for the pit, and the larger areas needed for the waste tips and tailings pond. But as we shall see in Chapter Four, the whole area of an openpit mining operation is far greater than the sum of its parts. We have also failed in this example to add in areas for the overburden tip, roads, ore-dressing plant, reservoir, workshops, offices, stores, labour camp, stockpile areas between various stages of milling, and so on. In practice, much of the overburden and some waste would be used, if suitable, in building the tailings dam. Sometimes tailings are made to form their own dam, though escaping silt is then a common problem.

Here is another sample calculation with a slightly different thrust. Suppose we start with 40 million tons of porphyry copper grading 0.81 per cent Cu plus – as a pleasant bonus – a few ppm Au and a few dozen ppm Ag. The same orebody might also be described as, for example, 49 million tons grading 0.68 per cent Cu at 0.20 per cent cut-off, or as 33 million tons grading 0.92 per cent Cu at 0.40 per cent cut-off, plus the gold and silver in each case. This is a relatively small orebody, but it could well be workable in an open pit with, say, a £30 to £35 million investment (yielding a £50 to £60 million return). We might find that the best working life might be twelve years at a throughput just under 10,000 tons of ore per day – fairly small for this type of operation, but the gold and silver values will help to make it payable. We can compute that this throughput corresponds to working rather more than 3 million tons of ore a year; at a moderately high waste/ore of, say, 3, the total waste produced will be 3 × 40 million tons = 120 million tons, while the total copper production (at 85 per cent recovery) will be 40 million tons × 0.0081 × 0.85 = 275,000 tons Cu in concentrate, or 23,000 tons a year. We might fairly assume that the ore-dressing plant in a mine of this size would work twenty-four hours a day for about 360 to 365 days a year, and the pit in two or three shifts a day for perhaps 250 to 300 days a year.

We would probably not be too far off if we surmised that the pit, with 45° average slope to its sides in rock averaging 2.8 ton/m^3, would cover (for example) about 1060 × 530 m (3480 × 1740 ft) at a maximum depth of 250 m (820 ft) (see below for details of calculation). There would thus be about 25 benches, each 10 m high, and the daily ore blocks would be about 55 × 6 × 10 m (3300 m^3, or about 9200 tons). Initial site preparation – making roads and ramps to provide access for stripping overburden – might easily entail moving some 3 or 4 million tons of soil and rock. The overburden itself might cover the pit area about 10 m deep and have a density of about 2.3 – a total of about 13 million tons, to be removed over a year or two at a cost of perhaps £1.5 to £2 million. (Our earlier lead–zinc example, by the way, neglected all this initial engineering.)

We might find that moving the amounts of rock required by our rough mining estimates might require about 30 m^3 of loading capacity, spread amongst several shovels for flexibility, plus a fleet of a couple of dozen 45-ton ore trucks or – perhaps more cheaply – about half as many twice as big. Loading ore[9] might account for nearly one-fifth of the pit's running costs, and hauling for more than two-fifths. We might need perhaps four 4½- or 6-inch drills for blasting. This and other capital equipment might cost us £5 to £8 million over the life of the mine. We would need a great deal of support equipment – for example, somewhere to store a few hundred thousand litres of diesel fuel, a source of 10 tons of mill steel a day and of four times as much lime (we would probably have to open our own limestone quarry and kiln), a reservoir supplying an uninterrupted 10 tons of water per minute, a power-line to bring in 20 million watts of electricity (half of it for the ore-dressing plant), perhaps £1 million worth of houses, an explosives magazine, warehouses, offices, parking areas, laboratories, garages, a changing-house, a canteen, machine shops, a welding shop, a tool shop, service centres supplying petrol and diesel fuel and water and tyres, radios to coordinate operations and a computer to plan them.

The mine, perhaps fortunately, is too small to support a local reverberatory smelter, which requires at least 40,000 or 50,000 tons of Cu in concentrate per year, though a new type of smelter might perhaps pay at a mere 30,000. We would therefore probably need a way to ship out a few hundred tons of concentrate each day to a remote smelter. Rail, conveyor, and pipeline are generally economic only for large outputs – well over a quarter of a million tons of concentrate per year – so we would be reduced to trucks, or to a short and uneconomic pipeline to a railhead. We would need more than a km^2 ($\frac{2}{5}$ square mile) for a waste dump and about twice as much for a tailings pond. We could not cut hauling costs by putting the mill very close to the pit, lest it be damaged by shock-waves from blasting or by subsidence of the pit walls. (The entire mill of a South African mine once vanished into a 500-ft-deep cavern hollowed out by escaping water.) We must be certain, of course, that our plant is not built on top of an undiscovered orebody! If we were short of money, we would have failed long ago: exploration and design studies probably cost some £1 million up to the decision to open a mine, plus at least as much on design after that. Just designing the £10 to £15 million ore-dressing plant probably took a year and cost over £100,000.

Finally, Table 8 summarizes some characteristics – including pit dimensions – of three sizes (small, medium, and large) of openpit copper-mines with stated outputs. All formulae are given in case you need them to check these calculations as an exercise. No non-ferrous metal is being mined in Britain in an open pit as big as the smallest of the three; on the Continent there are at least four such mines[11] each producing more than 3 million tons of ore a year, and in the USA there are dozens.

So far we have been estimating mine sizes from known grades and reserves. But what if all you know is that someone is evaluating a disseminated copper deposit? What you can do then is to 'guesstimate' the size of a minimum economic openpit mine. Though it is hard to generalize about the complex ore families such as lead-zinc-silver, for copper there are some easy rules of thumb. For example, to mine a low-grade copper deposit with present technology and zero profit, the price of copper (in 1973 £/ton) times the grade (as a percentage) must be at least 190; for an acceptable profit, about 230. This is obviously a crude rule, for it completely neglects such matters as overburden thickness and waste/ore; these should respectively be less than 50 m and less than 5 – preferably much less. As for reserves, these should suffice to produce 20,000 to 25,000 tons of copper a year in concentrate for, say, twelve years – that is, the orebody should contain about a quarter of a million tons of copper. This might be, for example, 50 million tons at 0.5 per cent or 25 million at 1 per cent: disseminated deposits are rarely richer than that. Of course, poorer deposits may be payable if 'sweetened' with a bit of gold or silver or molybdenum[12], or if workable on a big enough scale – a deposit grading less than 0.40 per cent might be payable if the reserves could support a throughput of tens of thousands of tons per day. But that would be irrelevant to the notion of a minimum economic openpit mine – a notion that requires us to assume a small, relatively rich disseminated deposit.

Disseminated sulphide ores of nickel are now workable in open pits at grades above 0.60 per cent or so, disseminated tin ores from about 0.2 per cent, and disseminated tungsten ores from perhaps 0.3 per cent or

TABLE **8**

SAMPLE ESTIMATES FOR THREE OPENPIT COPPER-MINES

Assumptions

Grade	G = 0.0065 = 0.65% Cu	
Waste-to-ore ratio	W = 2.0	
Concentrator recovery	R = 0.85 = 85%	
Concentrate grade	C = 0.25 = 25% Cu	
Mine life	Y = 14 yr	
Bank-to-waste swell factor	S = 1.75	
Bank-to-tailings swell factor	T = 2.40	
Average ore or waste density	D = 2.7 ton/m³	
Overburden average depth	B = 10 m	
Overburden swell factor	A = 1.20	
Power for ore-dressing plant	P = 17.6 kW-hr/ton	
Total power for mine	E = 3 × P	
Makeup water	F = 2 ton/ton ore	
Mill steel consumption	M = 1 kg/ton ore	
Lime consumption	L = 4 kg/ton ore	

All ore in place is assumed to be recoverable.

The pit is assumed rectangular in plan with 45° average slope. Volume is thus depth × (length — depth) × (width — depth).

The pit is assumed to run 300 days a year; the plant, 360 for 24 hours a day.

The tip is assumed to have maximum height 100 m and average height 67 m; area is thus 50% greater than for a cube 100 m high.

Quantity	*Unit*	*How computed*	*Results (in round numbers)*		
a Assumed Cu in concentrate	ton/yr		25,000	60,000	100,000
b Cu input to concentrator	ton/yr	*a*/R	29,400	70,600	117,600
c Ore input to concentrator	ton/yr	*b*/G	4,525,000	10,860,000	18,100,000
d Throughput of concentrator	ton/day	*c*/360	12,600	30,200	50,300
e Total ore mined	ton	*c* × Y	63,350,000	152,000,000	253,400,000
f Total waste mined	ton	*e* × W	126,700,000	304,000,000	506,800,000
g Total rock mined	ton	*e* + *f*	190,050,000	456,000,000	760,200,000
h Pit production	ton/day	*g*/300Y	45,200	108,500	181,000
i Total rock mined (bank)	m³	*g*/D	70,390,000	168,900,000	281,500,000

j	Pit area . . .	m^2	*	655,000	1,156,000	1,636,000
		acre	*j*/4047	162	286	404
	(or size for length = 2 × width)	m × m	*	(572 × 1145)	(760 × 1520)	(904 × 1809)
k	. . . for maximum pit depth	m	*	200	320	430
n	Volume of orebody if compact	m^3	*e*/D	22,060,000	56,300,000	93,850,000
o	Diameter of cylinder *k* m high to have volume *n*	m	$1.13 \times \sqrt{(n/k)}$	387	474	528
p	Overburden (bank)	m^3	*j* × B	6,550,000	11,560,000	16,360,000
q	Overburden (loose)	m^3	*p* × A	7,860,000	13,880,000	19,640,000
r	Waste (bank)	m^3	*f*/D	46,900,000	112,600,000	187,700,000
s	Waste (loose)	m^3	*r* × S	82,200,000	197,000,000	328,500,000
t	Total tip volume	m^3	*q* + *s*	90,160,000	210,900,000	348,100,000
u	Tip area [see assumption]	m^2	*t*/67	1,352,000	3,164,000	5,220,000
v	Total concentrate output	ton	*a* × Y/C	1,400,000	3,360,000	5,600,000
w	Concentrate shipping rate	ton/day	*v*/360Y	278	667	1,111
y	Total tailings	ton	*e* — *v*	61,950,000	148,700,000	247,800,000
z	Total tailings	m^3	*y* × T/D	55,000,000	132,200,000	220,300,000
	Area of tailings pond for average depth 20 m	m^2	*z*/20	2,750,000	6,610,000	11,000,000
	Power for entire mine	kW	*d* × *E*/24	27,700	66,400	117,000
	Mill steel consumption	ton/day	*d* × M/1000	13	30	50
	Lime consumption	ton/day	*d* × L/1000	50	121	201
	Clearwater consumption	m^3/day	*d* × F	25,200	60,300	101,000
	Area of 100-day clearwater reservoir for average depth 20 m	m^2	*d* × F × 5	1,260,000	3,020,000	5,030,000
	Probable minimum site	m^2	[approx.]	12,000,000	30,000,000	50,000,000
	Representative capital cost	£	[approx.]	35,000,000	90,000,000	150,000,000
	Representative labour force	men	[approx.]	170	400	800

* For given volume *i*, assumed maximum depth *k*, 45° sides, and length = 2 × width, $i/k = 2(\text{width})^2 - 3k(\text{width}) + k^2$ [from volume formula above]; solve for width by algebra or (more easily) by substitution.

N.B. Obviously not all the digits given in the table are significant, since the initial data often have only one or two significant digits.

less; but all such operations would have to have high throughput and large area – tin–tungsten mines, a common combination, tending to be perhaps smaller than copper mines. The best guide to what can be economically mined is the experience distilled in current mining journals and in mining companies' brochures and reports. The state of the art expressed in those sources is a foundation being continuously built upon by persistence and ingenuity.

Part Four: Finance, Law, and Politics

So far we have not put clearly in perspective the times needed for the various phases of mine development. From the start of exploration to the start of production generally takes between six and twelve years, averaging about nine; which end of the range one is closer to depends on the depth of the orebody, the remoteness of the site, and the difficulty of design. Even in the jungle-choked cliffs of Bougainville, from first prospecting to first production took only about eight years. Ordinarily about four or five years will be taken in exploration, evaluation, and feasibility studies; a further one or two in design; and, once the money has been raised, about two in construction of roads, foundations, ore-dressing plant, and utilities, and in pre-stripping.

Once capital has been committed to the project – once construction starts – all work must proceed with the greatest possible speed: with each day's delay, interest charges consume potential profits.

These times make no special allowance for getting any permissions; this process[11,13–15] could conceivably delay a mine by several years or prevent it altogether. Probably no really determined attempt to stop a mine has occurred in Western Europe in recent years.

Mining on this scale may involve an investment of hundreds of millions of pounds – more than the annual GNP of a small country. Few mining companies can afford this themselves; most must borrow from a bank or a consortium of banks, generally using as collateral the contracts for the sale of the mine's output over most or all of its life. The ease of borrowing may depend on the ability of the company to trade favours, not merely to remain profitable; for example, it is common to seek finance in the countries where the concentrates will be smelted and refined. Another common arrangement is to form a special holding company in partnership with other resource companies in one's own country or overseas, so as to combine both expertise and capital to mutual advantage. All such arrangements are sensitive to nuances of world politics and to jittery exchange rates. They are also, on occasion, shrouded in obscurity, since public knowledge of certain alliances might later prove an embarrassment. Competition among the world's mining giants – the companies with the human and fiscal resources to undertake large-scale openpit mining – is often less real than fictitious; some mining houses that appear to be rivals are happily helping each other, in various joint ventures, to divide mines and markets neatly between themselves.

Another more or less covert form of mutual back-scratching goes on between national or local governments and mining companies anxious to set up shop locally, often through a specially created local subsidiary with a local name. The company may want a tax holiday or other incentives; the government may want guaranteed employment, or spinoff enterprises such as commitments to build new smelter capacity or to buy mining equipment from domestic manufacturers. Such wishes generally have a way of being met, though regrettably rarely in the public view. Over the years, cordial working relations between industry lobbyists and regulatory civil servants have in many countries ripened into intimacy. Even closer contact will result from the administration of, for example, the UK's £50-million exploration/evaluation grants.

Many large mining companies have become fully integrated resource companies. The same group may now make the explosives and trucks, design and operate the mine, restore the land, smelt the ore, refine and semifabricate – even fabricate – the metals, and do all the marketing and administration, all through various subsidiaries that help to own each other. Profits can then be kept in the family and accounts adjusted to minimize tax.

The legal framework of mineral exploration, evaluation, and mining varies greatly from one country to another. In Britain, where the rights to metalliferous ores except gold and silver are in private, not State, ownership and may be severed from ownership of the surface rights, the position is far too complex to treat in this space, though it is thoroughly discussed elsewhere[13,16,17] in sources which you should consult if faced by an actual proposal. Further complications arise because mineral operations are aided by various statutes and controlled – in theory, anyhow – by the Town and Country Planning Acts[13,16] and by many other statutes[16] – among them the Alkali Acts, the Rivers (Prevention of Pollution) Acts, the Water Resources Act, the Deposit of Poisonous Waste Act, and others more obscure.

Under existing law[16], the citizen's main method of opposing proposed mining is to present his views at a Public Inquiry under the Planning Acts, if the Minister should care to hold one. I consider here only the case of opposition, not out of prejudice, but because mining companies are well able to look after their own interests. Unfortunately, the Inquiry procedure is so defective[13–15] that it is often not much better than an elaborate farce; and final, unreviewable discretion rests in each case with a local planning authority or with a Minister, who may consider or ignore what evidence he likes, even assuming it to be true. It need not be true or complete unless the applicant so desires. Assurances given at Public Inquiries are not binding. The only source of technical information on the applicant's intentions is the applicant, who can reveal or conceal whatever he wishes. Enforcement of planning control is discretionary, laborious, and often ineffectual – a fair description of the whole system[13].

Exploratory drilling needs planning consent; but because the law seems clear to everyone outside the mining industry, there is no case-law exactly in point, and until there is, mining companies are free to claim that drilling does not need consent. They may drill without it if neither the landowner nor the authority cares to stop them. Even if consent is sought, the application is likely to mention only a limited number of boreholes, and the authority or Minister is likely to consider possible future mining immaterial. Permission can thus be granted on the instalment plan, without ever examining the project as a whole before it gains immense economic and political momentum[18]. Short of years of risky litigation to overturn ill-judged precedents, there is no remedy but to seek better laws. Many substantive suggestions for reform[13] have been made to the Stevens Committee set up by the Government in August 1972 and due to report in 1974 on the interface between planning law and mineral operations.

A private mining company which can satisfy the Minister that there are any of certain common minerals under your land, which could be worked but for your refusal to agree to reasonable terms, can get compulsory mining rights from the High Court under whatever terms the Court may fix. If your land is scheduled in the local Development Plan for any sort of mining, the Court must assume that the mining proposed by the company is in the national interest; if your land is not so scheduled, the company must prove national interest in an unspecified manner[13]. In theory, the same rather cumbersome procedure, rarely used, could also yield compulsory exploration rights, but it is hard to see how a company could prove the existence of minerals for which it was seeking leave to explore[19]. Curiously, mining companies need not have got involved in UK exploration at all, for there is a government agency – the Institute of Geological Sciences – specifically charged with this task. For some reason, the poorly funded IGS did not share in the £50 million grants.

Landowners not anxious to help prospectors should note that the technical effectiveness of a systematic regional mineral survey is substantially reduced by the refusal of even one landowner to allow prospecting on his land; and he cannot in practice be compelled to allow it if he does not want to. Threats of compulsory exploration powers are largely hot air; at least, they are until the mining companies succeed in obtaining new legislation that has been in preparation for several years. Even aerial mapping and exploration[13,15] may ground an action in trespass or nuisance: the Civil Aviation Act 1949 frees overflying for transport from common-law liability, but prospecting is not transport.

Bearing in mind the importance of your land to the thoroughness of a regional survey, you are in a position, should you wish to strike a bargain with a mining company for exploration rights, to hold out for a high price, rather than the usual nominal fee. They are taking a business risk and must be prepared to find it expensive. What you have to sell they can buy only from you. They are likely to tell you that the survey is of national strategic importance – part of a national programme of proving mineral reserves. That is perhaps a side-effect, but hardly a motive: the exploration is purely a private business investment. Were it otherwise, the Government should fully control it or undertake it themselves.

Prospecting is normally done under a one- or

two-year option agreement, which should, but does not always, lay on the prospector a duty of care and of making good afterwards, besides putting right any unexpected damage. The agreement should not be assignable unless you want it bought and sold by a firm of mineral agents. Exploration options often include a first option on the right to extract – a very tricky matter calling for prior professional advice. For example, the agreement may bind you to offer the mining rights to the option-holder at any price anyone else offers – perhaps a penny, for example, offered by his cousin.

The mining industry moves in a cloud of habitual and largely unnecessary secrecy. Secrecy is perhaps legitimate until needed options have been obtained, but thereafter is hard to justify. Nonetheless, normal commercial practice in managing a prospect seems to be that the inhabitants are the last to know what is happening[18]. If you fancy you see little men tapping at rocks in your back garden, perhaps your paranoia is just heightened awareness.

Because openpit mining extinguishes surface rights by extinguishing the surface, mining companies usually buy the surface freehold. They either buy the mineral rights or lease them under a 'preference partnership' with the owner, whereby he gets a royalty based on gross revenues before deducting any costs. Royalty negotiations may be lubricated by the existence of the High Court compulsory mining powers.

If a mine amounts to an actionable nuisance – for example, by reason of dust, noise, or other pollution – damages or injunctive relief may be available. If you lose such an action, you may have to pay the costs of both sides, Legal Aid being rarely available in nuisance actions, and if you win, you may still be out of pocket. In an unusually clear case, you might be able to get an injunction against an adjacent mine before it even opened. For this reason, the proprietors of a large mine are likely to make every effort in advance, if you live next door, to buy either your land or you.

Notes

1. L. B. Slichter: The need of a new philosophy of prospecting, *Min. Engineering*, NY, 12:570 (1960).
2. D. A. O. Morgan: A look at the economics of mineral exploration, Ninth Commonwealth Mining and Metallurgical Congress, 1969.
3. D. R. Derry: Man and minerals . . . exploration, *Bull. Can. Inst. Min. Metall.*, 80:107 (April 1959). Compare A. A. Brant: The pre-evaluation of the possible profitability of exploration prospects, *Mineralium deposita* 3:1 (1968).
4. H. G. Ewanchuk: Open-pit mining at Bethlehem Copper Corporation, Ltd. . . ., Ninth Commonwealth Mining and Metallurgical Congress, 1969. The Bethlehem operation has recently been much enlarged by extension to include new orebodies.
5. Not all copper deposits are disseminated; some, geologically different and richer (1 to 3 per cent or more), are well-defined, often lens-like, with a compact shape generally controlled by geological features. Well-defined copper deposits may be up to about 50 m thick, and may be folded. Layers

tens of m thick are surface- or underground-mined according to depth; the latter technique might produce only a tenth as much waste and tailings as would an openpit mine producing the same yield from disseminated ore. It is therefore important to know whether a particular copper deposit is disseminated or well-defined. A geologist can normally get a good idea of this by inspecting and interpreting the site.

6. Technologic Trends in the Mineral Industries (Metals and Nonmetals Except Fuels), *Minerals Yearbook 1970*, US Bureau of Mines.
7. J. D. Lowell: Jacklin Award Lecture, *Min. Engineering*, April 1970.
8. See Introduction, note 3.
9. T. Atkinson: Selection of open-pit excavating and loading equipment, *Trans. Instn. Min. Metall.*, 80:A101 (1971).
10. In soft rock, ripping tools, rapid temperature changes, high-pressure water-jets, or mechanical shovels may be substituted for blasting, but such rock is rarely the subject of deep openpit metal-mining. The machines used in a soft-rock surface mine may be very large: one US walking dragline, for example, has a 220-yd^3 (170-m^3) bucket on a 310-ft (95-m) boom, weighs 12,700 tons, and draws 36,000 kilowatts (48,500 hp). Such devices, like multi-bucket chain or wheel excavators, are often used for stripping coal, perhaps supplemented by giant augers.
11. 'Commission' on Mining and the Environment (Lord Zuckerman, Chn): *Report*, London, 1972.

12. A remarkable example, described on pp 416–423 of the June 1972 *Mining Magazine*, is the new Brenda openpit Cu–Mo mine in British Columbia, whose 177-million-ton reserves grade at only *0.183* per cent Cu but are sweetened by a large 0.049 per cent Mo content. The orebody, some 2800 × 1300 ft in plan and 900 ft thick, has a richer 26-million-ton core (0.212 per cent Cu, 0.063 per cent Mo) which will be worked during the initial three-year tax holiday. Low waste/ore (about 0.9) and high recovery (initially 88 per cent Cu, 82 per cent Mo into separate concentrates grading 21 per cent and 56 per cent respectively) help to make the mine economic over a 21-yr lifetime at a throughput of 20,000 to 24,000 ton/day. At the end of 1971 there were 304 hourly-rated and 91 salaried employees. The high capital cost – £1950 per ton/yr of Cu output capacity – is made up for by the valuable Mo output, even though the Mo concentrate has only about one-tenth the tonnage of the Cu concentrate. Other details include: primary crusher below the level of the pit rim so that initial hauling is downhill; drills 12¼ inches (31 cm) in diameter; explosive charge to shatter the hard granite, about 0.5 pound (230 g) per ton; working (not including standby) machinery, two 11-yd^3 (8½-m^3) shovels and eight 100-ton trucks.
13. Friends of the Earth Ltd: *Evidence Submitted to the Committee on Minerals Planning Control* (Sir Roger Stevens, Chairman), London, 1973, including Addendum thereto.

14. Friends of the Earth Ltd: Rock-Bottom: Nearing the Limits of Metal-Mining in Britain (evidence submitted to the 'Commission' on Mining and the Environment); reprinted *Ecologist*, May 1972, p 4.
15. A. B. Lovins and P. H. Evans: *Eryri, the Mountains of Longing*, Friends of the Earth Inc with McCall (Saturday Review Press, New York), 1971, and with George Allen & Unwin (London), 1972.
16. O. Thorold, *The Environmental Law Handbook*, Earth Island (London), 1973 (in press).
17. J. A. Williams: Proposed changes in mineral legislation in the United Kingdom, *Trans. Instn. Min. Metall.*, 80:A38 (1971).
18. Do You Dig National Parks? (Michael Barnes, Producer), extended *Horizon*, BBC-TV, 22 May 1972; film copies available from Friends of the Earth Ltd, 9 Poland St, London W1, and from Concord Films Council, Nacton, Ipswich, Suffolk.
19. It also appears[13] that a mineral undertaker whose relations with the local authority or the Minister are sufficiently cordial can get compulsory rights without going to the High Court at all.

4

DURING MINING

In discussing openpit mine design, we have already touched on some of the operations affecting life and land at the site and nearby. This chapter will add further practical details.

Physical disturbance is likely to be fastest and social disruption greatest during the phase of construction and pre-stripping. This work entails very heavy traffic flows: a large labour force (Chapter Six) and all the heavy machinery and materials, such as concrete for foundations and parts for the ore-dressing plant, must be brought in. Noise and dust will be similar to what might result from the simultaneous construction of a major motorway junction and a steel-mill, both on the same site. Also, from this phase onwards, normal water-flows through the site will be diverted or dammed. The removal of all vegetation from working areas will increase run-off, perhaps presenting a danger of down-stream flooding or erosion.

Modern technology can change landscape with remarkable speed. At Bougainville, perhaps the best example, a thick tropical jungle was first killed with herbicides and then cleared by blasting and high-lead logging – drawing a steel hawser rapidly through the air between two powerful winches. The damp vegetation was then burned in heaps with the help of fuel oil and large fans. Some 28 million tons of soft overburden was blasted away with water-jets from 725-hp pumps, then bulldozed (at over 200 m^3/bulldozer-hour) and washed into a nearby mountain river – whose gradient, once it has received some 400 million tons of waste, will go from 8 to 1 per cent.

A major problem in working mines is to keep the pit from filling with seepage and rainwater, which makes loose rock heavier, bogs down equipment, and leaches out soluble metals such as copper and iron. The leaching of sulphide ores also produces, through oxidation to sulphates, a solution akin to dilute sulphuric acid. Such acid drainage must be neutralized with lime before it leaves the mine – if, that is, it can be captured.

Water entering the mine site, whether through rain and seepage or through pipes, can leave only in three ways: evaporation from the tailings pond, deliberate discharge into the watershed, and accidental discharge via run-off or seepage. The third method, and to some extent the second, inadequately controls the escape of toxic metals mobilized by digging up large amounts of mineralized rock and leaving them lying in the rain. Leaching can be reduced – for example, by storing ore stockpiles under cover at twice the cost – but not eliminated, if only because one cannot hold an umbrella over the whole site. Metals are, of course, being leached all the time – that is why geochemical prospecting works – but the rate of leaching increases enormously during mining. Despite much hand-waving by the mining industry and costly control efforts in some Canadian mines, no technology appears to exist for preventing the release of leached metals in potentially dangerous concentrations. The nearest approach is reducing the seepage rate from the pit by grouting fissures.

This escape of toxic metals[1] – lead, zinc, cadmium, mercury, copper, antimony, tellurium, selenium, arsenic, nickel, chromium-6, radium, and others – is one of metal-mining's major unsolved problems. Toxic metals are poorly-understood cumulative poisons. Most are eliminated from the body very slowly. They can be concentrated in food chains, and are extremely persistent: many streams in old mining areas are lifeless because, a century or more after mining stopped, there is still enough leaching to provide (say) 0.5 ppm Cu or 0.1 ppm Cd. Sixty parts Cu per billion can be toxic to trout and other fish[2]. Escaping metals could more or less permanently sterilize a watershed including bottomland, a major river or estuary, or conceivably a substantial stretch of coastline. In a classic example, an attempt by the White Fish Authority to hatch oysters in the Conwy estuary of North Wales in 1966 failed because of metals in run-off from small and long-abandoned lead–zinc workings high above in the Carneddau[1,3].

Metal-bearing dust often escapes from the mill or by blowing off waste tips. Such dust can affect soil, wildlife, farming, and man up to about 8 km (5 miles) from an openpit mine in prolonged dry weather.

Some reagents used in concentration are malodorous, though this is rarely a problem outside the site. Water pollution by residual reagents is a more likely problem.

Blasting is typically done once or twice a day with a

minimum of about a ton of explosives per blast. The best type, amount, and placement of explosives cannot be exactly predicted, but must be worked out empirically. Blasting produces noise, shock waves in the ground, and dust which may drift downwind as much as a mile or two. The noise and vibration may be a nuisance for several miles, though the noise, from blasting as from the mill and machinery, could be reduced by baffles. Special mounds or ridges would be needed to baffle both sound and sight.

Both mill and machinery are extremely noisy – well above most modern noise standards[4–7] – and mill noise is especially noticeable at night. There have been no proper studies of mining noise, nor are there systematic procedures for limiting it.

It goes without saying that the mine site, which must be well fenced for security, is a less than ideal wildlife habitat, though the well-publicized hippos intrepidly wallowing in the Palabora tailings area might disagree.

As we saw in Chapter Three, the weight of concentrate leaving a mine may be readily estimated; so, therefore, may the necessary traffic, perhaps in 12- or 15-ton trucks if the concentrate must go to a port or railhead by truck rather than by a more expensive pipeline or conveyor. Transport of men and materials into the site may prove an equal problem, without any option but road traffic. Under present British law[7], licensed vehicles may use public roads as much as they like without bearing the cost of any necessary strengthening or repairs. Mud and spillage on roads are another common problem.

If a suitable dam site can be found, tailings are

usually pumped to and impounded in a nearby valley, the narrower the better. The tailings pond is not, as its name implies, a clearwater lake, but rather a large and probably lifeless mud-flat (Chapter Five). The dam is subject to normal engineering safety demands, and the pond itself, like the clearwater reservoir, must not by its weight or by seepage activate dormant geological faults.

Everything that enters the mine site must go somewhere. All reagents – including those poisonous synthetic oils and powerful frothing agents – are as far as possible recovered, for they are costly, or they are lost. Even the more innocuous tens of tons of lime a day must go somewhere – generally downstream. All the fuel is burned, producing air pollution of one sort or another. All the lubricating grease, the sewage, the cast-off bits of machinery replaced by spares, the rubber that used to be on those worn-out tyres – all of it shows up somewhere sooner or later. This 'where-does-it-go?' approach is useful in tracking down subtle losses and emissions. It is especially helpful in dealing with a gold or silver concentrator, which will normally use large amounts of sodium cyanide – very nasty if it gets out.

Knowing the likely emissions may not do you much good. Under present British law[7], air and water pollution are controlled by special agencies whose dealings with the mining company would be secret: indeed, the penalty for polluting a river (on first offence) is less severe than that for saying who is putting what into a river. The actual monitoring of emissions would normally be done by the mining company itself, and the results would be secret. These secrecy provisions are unnecessary and long overdue for repeal; but while they are still law, mining companies rarely find them inconvenient.

Once begun, openpit mines tend to expand to include new orebodies nearby. The economic life of the very costly ore-dressing plant can thus be extended. But even where this is not done, openpit mines tend to sprawl over surprisingly large areas, especially in hilly country where topography constrains design. The tailings pond, linked to the rest only by a pipeline and by the need to haul dam material, can be 10 miles (16 km) or more away if topography so dictates. With a standard layout – pit, plant, waste tips, and tailings pond as close together as convenient in mildly rolling country – the total site area may well be fifteen or more times the area of the pit, which in turn is at least twice the area of the orebody. The total site area may be three or more times the combined area of the main parts, since these parts cannot be crowded, but tend to radiate rather like a starfish, thus affecting more adjoining land. For example, the 1969 Bethlehem mine, which is not much above minimum economic size for copper, directly and drastically affects some 3000 to 3500 acres (1200 to 1400 hectares or 4.7 to 5.5 square miles). At the other extreme, the lease area of the Bougainville mine (Table 6) – the largest in the world in start-up tonnages – is about 23,000 acres (9300 hectares, 36 square miles).

On any conventional design criteria, the total site area of a minimum economic mine for disseminated copper in typical British conditions would be about

4 to 8 square miles (1000 to 2000 hectares), all centred round a pit of perhaps 100 to 200 acres (40 to 80 hectares). Complex or heavily faulted orebodies are sometimes mined in several small pits rather than one large pit, and with new and untried techniques, such as progressively putting the waste of one pit into another, the total area might conceivably be slightly reduced; but the main problem would still be the disposal of waste and tailings that had swelled far beyond their original volume and would not fit back where they came from.

This 4- to 8-square-mile minimum estimate includes a narrow controlled-access corridor within the lease area, but does not take account of adjacent land indirectly affected by the mine, whether by noise, dust, vibration, or run-off or by subsidence sometimes caused, especially in soft rock, when a large pit disturbs the underground flow of water.

Openpit metal-mines are sometimes only 100 m or so (330 ft) deep, but for such steeply dipping orebodies as a porphyry-copper 'pipe' the pit may be 300 to 500 m deep (1000 to 1600 ft). Such deep evaluation drilling is seldom reliable, so evaluation of ore at depth may have to wait until the pit is largely worked out. It may therefore be wise to count on extra wastes and tailings in case of an underground continuation – or in case rising prices lower the cut-off grade during the working of the pit, thus effectively enlarging the orebody.

On the other hand, falling prices may indefinitely delay the start-up of the mine. It is also possible for a company to curtail expansion – or, more rarely, production – at a particular operating mine in order to employ its capital more profitably elsewhere.

British planning law makes provision for opening mines but not for closing them: they have an infinite theoretical life[8]. Besides biological and physical problems of restoration (Chapter Five), local authorities thus have the further difficulty that they can never compel restoration. The company can say – perhaps truly – that it has only paused, has not bottomed the deposit, and plans to continue when prices rise.

Notes

1. See Chapter 2, note 10.
2. M. Ellis: Detection and Measurement of Stream Pollution, *Bull. US Bur. Fish.*, 48:365 (1937).
3. H. Elderfield *et al.*: Heavy metals and oyster culture in Wales, *Marine Poll. Bull.*, 2:3 (March 1971).
4. [Wilson] Committee on the Problem of Noise: *Report*, HMSO, London, 1963.
5. Working Group on the Noise Abatement Act: *Neighbourhood Noise*, HMSO, London, 1971.
6. Method of rating industrial noise affecting mixed residential and industrial areas, British Standard 4142, 1967.
7. See Chapter 3, note 16.
8. See Chapter 3, note 13.

5

AFTER MINING

Whenever mining stops – 'cessation' is legally undefined – what is to be done with the mess? The operator may wish to do nothing – not just to save money but because, with rising prices, his waste tips and tailings, and even his waste below cut-off still unmined in the pit, may become payable, so it may be premature to backfill the pit and dismantle everything. But let us suppose that the company is able and willing to put things right – a necessary pair of assumptions, since present law[1,2] cannot compel an unwilling operator to restore nor to insure or post bond against his insolvency. Let us further suppose that the mine has from the outset been designed, built, and run according to a restoration plan – otherwise little can be done. To what extent, then, can the buildings, pit, tailings pond, and tips be

restored, *i.e.* returned to their original use and general state? or

rehabilitated, *i.e.* put to another use, such as a pump storage scheme? or

reclaimed, *i.e.* changed from originally derelict land into productive land?

Buildings are easiest: they can in principle be dismantled in a year or two. But it is more plausible that a heavy industry would be sought to take over the site, buildings, and utilities in an attempt to profit from dereliction and ease the job decline (Chapter Six).

The pit is a much more intractable problem. It is very big. Filling it would cost at least as much as digging it, even if the material removed would fit back again, which it will not. The sides of the pit may prove unsafe unless extra money has been spent to make them stable by reducing their slope and increasing the waste/ore ratio: ordinarily, the ideal slope is one which collapses of its own accord as soon as mining has finished. Long-term safety would demand periodic maintenance in perpetuity. Benches could, of course, be blasted to a uniform slope to improve appearance, and for a price the lowest levels of the pit could be backfilled.

Possible uses for open pits have strained many imaginations[3,4]: ideas include flooding to form a recreational lake, part of a pump storage scheme, or a freshwater reservoir (for those who don't mind a bit of metal in their water); rubbish-tipping; or conversion to an amphitheatre. Uses that require a dry pit are probably fanciful; those that require a flooded pit may fall foul of permeable rock, stagnant water, or soluble materials. In any event, the use must be pre-planned, and seepage and run-off from the pit must be controlled for as long as they are toxic – perhaps a few centuries or so.

Tailings present other problems. If dry, they erode and blow away unless stabilizing grasses can be grown on them: attempts to do this abroad have met with mixed success[5]. If wet – very likely in Britain – or if too fine to drain internally, they will stay a soft silty mud, perhaps with a surface crust, indefinitely. The dam must therefore remain sound indefinitely in order to prevent an avalanche of toxic mud, potentially more dangerous than one of water. Present British law does not ensure adequate design and perpetual maintenance of tailings dams[6]. Furthermore, the water running into and falling on the tailings pond will leach residual metal values from the tailings and must therefore not be allowed to seep or flow away bearing toxic metals; it is not clear how such escape can be prevented, though vegetation would reduce it somewhat if any would grow[7].

Finally, waste tips, which cost too much to move very far, are also toxic. If a special mutant grass or shrub can be made to grow on waste or tailings[7,8], the metals will tend to enter the plant, which may then also be toxic. A further problem is tip stability, especially in hilly and rainy country. Tips can give way either suddenly in a liquefaction slide – as at Aberfan in Wales and Jupille in Belgium – if the tip becomes saturated with water, or slowly in a gradual creeping if gravity exceeds the shear strength of the foundation or fill or both. The first kind of slide can occur in unconsolidated waste, and tips in modern mines are therefore carefully compacted as they are laid down. The second kind of slide can in principle be prevented by proper design, but this may mean severely limiting slopes: the Tennessee Valley Authority considers slopes steeper than 28° 'unmineable' for coal for this reason, and the limit may soon be reduced to 24° or even 18°. The exact limit is, however, a matter for calculation and trial in each case.

Despite some pious optimism in the mining literature, the fact is that no large openpit copper-mine has ever been restored anywhere in the world. Methods of doing so are therefore hypothetical, and should be proven first on existing metal-mine dereliction. For example, an intending British copper-miner could be paid to establish a stable ecosystem in the lunar landscape of the old Parys mine in Anglesey – small by modern standards, but once the world's largest. This task may well prove impossible. Soil has a stable layered structure and a physical, chemical, and microbiological composition evolved through aeons of weathering, while mining wastes are sterile, toxic, structureless, and unorganized. Saving and returning topsoil, which is done successfully in progressive restoration in stripping coal and ironstone, may not work so well if the topsoil is fifteen years old. Besides, toxic metals will slowly leach outwards into any fresh soil cover; acidic products can be neutralized only in a superficial surface layer that gradually turns acid again; and nitrate fertilizers may help to remobilize bound metal ions. All this means that permanently stable vegetation on mining wastes is unlikely except in nature's own good time.

The prices of straightforward landscaping are also high. Nearly all the published data are for opencast coal-mining – shallow, progressively restored, and non-toxic, though there is often acid drainage. For what it is worth, simple grading to control erosion and allow planting seems to cost £200 to £2000 and up per hectare; simple grass/legume and tree planting, about £225 on US average; hydro-seeding – spraying seeds, fertilizer, and binder together, a technique which could probably grow grass on concrete – about £360, or another £80 with trees; landscape planting for recreation, over £1200. British costs for reclaiming coal spoil to parkland or farmland – doing a much better job – generally run about £5600 to £5800 per hectare, excluding overheads, and reportedly yield fertility roughly half to two-thirds that of top-quality farmland. All such techniques use a great deal of fertilizer, much of which runs off into the watershed and reappears in our drinking-water.

Such costs are a heavy burden. Some mining companies would like them paid by the public by way of an amenity grant; others will settle for a hidden subsidy in the form of tax relief on the restorative budget; all would like their responsibilities limited to the 'practicable' or the 'reasonably economic'. The public may have higher standards, and may well be wary of industry's statements[4] that 'demand for a redevelopment scheme should be determined toward the end of the mine's life. . . . [I]f assessment shows little demand . . ., the dumps can be revegetated and abandoned to blend into the landscape.'

Notes

1. See Chapter 3, note 13.
2. See Chapter 3, note 14.
3. See Chapter 3, note 17.
4. L. W. Saperstein: The Potential for Reclamation or

Redevelopment of Open-pit Mines, AIME Environmental Quality Conference, Washington, DC, 7 to 9 June 1971.

5. J.-C. LeRoy and Herman Keller: How to reclaim mined areas, ponds, and dumps into valuable land, *World Mining* (UK edn), January 1972, p. 34.
6. Institution of Civil Engineers: *Reservoir Safety* (Report of the ad hoc Committee set up by the I.C.E. to submit proposals for a revision of the reservoirs (Safety Precautions) Act 1930), I.C.E. (London), 1966.
7. R. A. H. Smith and A. D. Bradshaw: Stabilization of toxic mine wastes by the use of tolerant plant populations, *Trans. Instn. Min. Metall.*, 81:A230 (1972); R. F. Nielson and H. B. Peterson, Treatment of Mine Tailings to Promote Vegetative Stabilization, Bulletin 485 (June 1972), Agricultural Experiment Station, Utah State University (Logan, Utah).
8. Bauxite stripping poses special restoration problems: it may yield an insect-breeding swamp too wet to revegetate. Extracting a ton of alumina from bauxite also yields about a m^3 of 'red mud', containing residual caustic soda that may percolate into water supplies. See *e.g.* the Western Australian Environmental Protection Authority's report: Alumina Refinery (Upper Swan) Project, 18 Feb. 1972.

6

SOCIAL IMPLICATIONS

Openpit metal-mining is a capital-intensive heavy industry of limited life. In Britain and elsewhere it is likely to be practised increasingly in rural areas, often beautiful ones, with high unemployment and chronic depopulation – frequently an after-effect of earlier labour-intensive mineral workings. Mining is inherently a boom-and-bust industry. In an area of tenacious marginal economy, it may destroy the basis of existing employment depending on permanently renewable resources: for example, farming and tourism. Its effect on a fragile rural society would be profound and generally harmful. Mines grow no second crop: the main harvest would be degraded or devastated land, unemployment probably higher than before, and both precedent and new pressure for importing further extractive heavy industry. The main economic benefits would go not to the workers or the area – mining companies are not social-service charities – but to the shareholders, as the Annual Reports of any large mining company will show.

The large flows of men into an openpit mining project during the construction phase, when employment may run at three to five times the stable level later, are socially disruptive and normally require a labour camp. Recent large civil-engineering projects –

the Trawsfynydd and Wylfa nuclear stations, the Stwlan pump storage scheme, and the Holyhead smelter, all in North Wales, or the Milford Haven and Dounreay projects, to mention a few – have shown that the local labour pool may be surprisingly unaffected by hundreds of new job vacancies. There would be substantial spin-off stimulation of the local economy, though the new income, unlike that from, say, tourism, would probably not be very well distributed, and the construction boom would be short-lived (Chapter Three, part 4).

The number and character of jobs available at a modern openpit metal-mine in normal operation is often misrepresented and misunderstood. Here, therefore, is a typical and relatively labour-intensive staff list for the mid-life of a relatively small openpit copper-mine (25,000 tons Cu per year worked in two shifts a day):

operators: excavators 12, ore trucks 50, drills 16, blasting crew 8
shop: mechanics 10, machinists 10, smiths 2, bit sharpeners 2, crane operators 4, electricians 10, others 8
technical staff: geologists and technicians 8, surveyors 4
drivers for support vehicles (tank and repair trucks, bulldozers, graders, etc.): 16
clerical and support staff: 12
managerial staff: 8.

Several important features of this labour force must be noted. First, the total labour force of this £30 to £40 million mine – slightly above minimum economic size – is roughly 180 or fewer, not 500 to 1000 as is sometimes presumed: mines of this sort are highly mechanized. The Bethlehem [Jersey and East Jersey] mine, slightly smaller and more sparing of manpower, employed[1] at its mid-life in 1969 only 23 salaried and 130 hourly-rated workers. Second, few of the jobs can be walked into; most are fairly skilled and some very highly skilled. Most of the jobs are therefore unsuited to the needs of rural areas of high unemployment, and indeed are commonly filled by contractors and by migrant specialists who go from one mine to another. This is not reversing depopulation so much as displacing or imposing on the residents. Third, most of the jobs are fairly strenuous and entail substantial physical labour on outdoor shiftwork; they are unsuitable for older or disabled workers, and thus cannot touch the hard core of rural unemployment. Fourth, local workers who do acquire the needed skills, perhaps through apprenticeship or a training programme, are unlikely to find local scope for their skills once the mine is worked out. In short, assuming that openpit mining can solve rural unemployment problems wholly ignores who is unemployed and why.

The extra jobs during the construction phase of two or three years – perhaps 300 to 600 for a small mine – will seldom be unskilled, but mainly machine-operating and skilled crafts such as would be needed to build a large factory to a tight budget and schedule. It rarely pays the company to recruit and train local plumbers, electricians, welders, riveters, bulldozer drivers, and so on when mobile skilled workers can be had at short notice from contractors in nearby cities. Those skilled and semi-skilled workers whom higher short-term wages lure from their local jobs may find no jobs to go back to – local industries often fail under the burden of such competition, and the local labour market tends to tighten as wage expectations rise.

If the mine is much larger than minimum size, its normal work force may be much larger too: a 50,000-to-60,000-tons-Cu/yr mine might employ 350 to 500, and a large mine (100,000 tons Cu/yr or more) 700 to 900 or more – rosters that on grounds of numbers alone, let alone skills, could hardly be filled from the local labour pool. But these mines would also devastate very large areas, say 10 to 30 square miles (25 to 80 km^2), and might therefore seem locally to be less of a bargain. The relatively labour-intensive style of openpit mining in poor countries, as at Palabora and Bougainville, is unlikely to be emulated in rich countries where wages are far higher and a skilled mobile labour force is readily available. In rich countries, every effort will be made to cut the labour force to a bare minimum.

When mining nears its end, with a gradual running-down over perhaps two to three years, goods and services the supply of which has expanded to meet increased demand will remain unsold; the so-called reverse multiplier effect, or backlash, may then produce local economic collapse with disastrous social effects. After-works programmes are unlikely to employ enough men to provide much of a cushion against this collapse. Many embittered mining and quarrying ghost-towns today show the poverty, high welfare burden, and bleak future that tend to follow in the wake of extractive heavy industry – even in its long-term, traditional, labour-intensive forms. Such consequences seem inherent in the nature of the industry.

The kind of industry that the economically depressed British countryside needs is not short-term and centralized but permanent and diffused – jobs which will offer enough security and advancement to persuade local school-leavers to stay in their villages. Ways to fill this need do exist[2], but a need from which modern openpit mining differs more, or to which it is more inimical, can hardly be imagined. Stable rural agriculture and tourism cannot possibly coexist with a large-scale mine. Nobody has yet done a thorough study of the local economic effects of such mining, taking into account the losses in income from agriculture, forestry, and tourism (or, say some mining companies, the gains from extra tourism). Regional tourism – as important a livelihood in parts of rural Britain as anywhere in the world – may suffer particularly from the loss of the calm, integrity, and solitude that most visitors seek and that residents most treasure when they have lost it.

Not merely the economic but also the social and cultural effects of large-scale mining must be soberly assessed before mining techniques quite new to Britain, as to most densely populated countries, are welcomed too uncritically. This is particularly true in hill-farming areas, where the relationship between man and land, ecology and economy, is most complex and delicate[2 3]. It is an important moral question, not just a political one, whether to reduce traditional cultural diversity by importing alien urban and industrial values into such places.

The only clear beneficiary of bringing a large mine into a depressed rural area is the mining company, whose grief at turning upside-down an area which, as a group of devout conservationists, it would prefer to leave untouched may be assuaged by various government grants and subsidies. By a curious coincidence, most of Britain's most beautiful places, which the government spends money to protect from unsuitable small-scale exploitation, are also classified as Development Areas in which a Minister may – partly at your expense – introduce any sort of heavy industry he likes, without any reference to Parliament. This is true even in Britain's National Parks, which, partly because they have no special legal protection, are not internationally recognized. The principles originally intended to govern mining in the Parks have apparently been abandoned[2,4]. The Parks' vulnerability to steady nibbling-away is heightened by their chaocratic form of administration[2] and by the failure of successive Governments to give them money. The Countryside Commission gets, for the annual upkeep of Parks covering 9 per cent of England and Wales, less than 1 per cent of what private mining companies have been given from public funds in order to help them to prospect, mainly in those same National Parks. These twisted priorities are rightly the subject of a growing public debate[2,4–6], fuelled partly by the recent publication of a detailed case-study of how a Park works, what is going wrong, and what could be done about it[2].

Notes

1. See Chapter 3, note 4.
2. See Chapter 3, note 15.
3. Sir Frank Fraser Darling: *West Highland Survey*, Oxford University Press, 1955.
4. See Chapter 3, note 14.
5. See Chapter 3, note 13.
6. See Chapter 3, note 18.

7

POLICY IMPLICATIONS

A work of this length can only mention some of the major policy issues raised by a proposal for an openpit mine, a peculiarly irreversible type of development; complete exposition of any particular case pro and con would fill many volumes. Mining companies naturally select those arguments which present their case most favourably. That they cite benefits to employment or to the national economy does not mean they act mainly out of altruism or patriotism – they act first for profit – but the arguments they adduce may confuse what is really at issue.

The central questions in a mining dispute should be: Is it necessary to mine at this time and place? If so, how can harmful effects be minimized? Or will they still be so great that mining should not be allowed?

In practice, however, arguments which should be peripheral, such as employment and balance of payments, tend to be advanced as central in order to support the case of the 'developer'. (A man taking money out of the bank would hardly be said to be 'developing' it, but such is the way of euphemism.) Were we seriously interested just in providing rural employment, we could build pyramids, perhaps out of compacted mining spoil from derelict land – a cheap, long-term, labour-intensive enterprise with low land needs and high tourist value, just the opposite of openpit mining in every respect. Likewise, we could undoubtedly help the balance of payments by shipping Westminster Abbey to Texas, block by block. If we are to distinguish such programmes from openpit mining, we must phrase arguments about 'necessity' with some care. Strictly construed, 'necessity' is never an issue, since no mineral in Britain cannot be bought on the world market. The argument tends instead to hinge on whether Britain or whatever country can afford to buy abroad and whether the supply is secure.

The first argument must depend on a detailed survey of the entire import-export position and history for the metal concerned and for relevant substitutes. Take care: metals travel hither and yon in dozens of different forms, and most trade statistics are inconsistent with themselves and with each other, requiring expert advice to interpret them properly.

It is misleading to use gross import figures alone. For example, Britain imports large amounts of gold and silver, and even diamonds, but then adds value – much less than the cost of the raw material – and re-exports. This is how a trading nation works: and improving even by a few per cent the efficiency with which resources are used may well save more foreign exchange than a domestic mine ever could, and more permanently too. More efficient recycling (Chapter Two) is a similarly painless and necessary way to reduce the need for mining.

Do not be lulled by large numbers. UK mineral production in 1970, for example, totalled about £1000 million, of which only £40 million was metals and only £4 million non-ferrous. That the UK imports some £600 to £800 million worth of non-ferrous metals a year does not mean that even, say, one-sixth as much could be readily saved by acceptable domestic mining. A Palabora in Britain, for example – perhaps an unlikely prospect both geologically and politically – would, during each year of its life, produce copper with a gross market value of only about £40 to £70 million – less dividends to foreign shareholders, losses to British shippers who now import copper, payments to heavy-equipment suppliers abroad, exports of capital for foreign re-investments, and so on. Indeed, demand is so large compared with possible domestic production that one medium-to-large British copper-mine, over its entire life, could produce only enough copper to meet British demand for some months – or world demand for a few weeks. Britain is unlikely ever to approach self-sufficiency in *any* non-ferrous metal of commercial importance, at least until depletion of foreign reserves enforces an involuntary and spartan self-sufficiency.

The argument that domestic production is essential to strategic security is not nearly as convincing now as it may be a few decades hence. If growing demand and dwindling supply will gradually raise prices and reduce suppliers' amiability, it would seem logical to conserve domestic supplies as long as possible – they will not run away – rather than burning our fuel before winter. With copper as with oil, the nation that selfishly values its own interests will buy abroad while it can, maintaining at home where possible a strategic reserve of the kind now required by the Forestry Act 1967 for timber, which is a renewable resource. Of course, this argument does not consider the morality of exploiting or polluting other countries for one's own uses. A mining company is unlikely to argue that it is time we tried on ourselves the sort of devastation which, perhaps through mines run by the same company, we have long been inflicting on others abroad. Yet just because the moral issue is not raised in practice, that does not resolve it or make it any less perplexing.

Foreign suppliers would lose whatever British market was filled by domestic production. If you accept the argument, used by multinational mining houses, that their activities bring only blessings to the countries concerned, then you might worry about withholding those blessings through competition. In the long run, though, foreign suppliers would have no trouble finding new markets.

There is a good case for the UK, like any trading nation, to diversify its sources of supply. But this has nothing to do with mining at home; it is more a matter of political expedience and habit. The oft-heard argument that domestic reserves are essential in time of war is not only a well-aimed boomerang if intended to support mining now, but is also out of date in an age when starting a mine may take two years and finishing a nuclear war two hours.

The balance-of-payments argument is probably the best argument for domestic mining, but it is not a very strong argument. A smallish copper-mine, for example, could produce a gross import saving of some £10 million a year for its life by inverting the centre of a county now earning more than £5 million a year from a grossly inefficient tourist industry; but this seems a Pyrrhic victory. On the national balance sheets, too, £10 million would hardly be noticed – it would suffice, for example, to keep the Concorde programme just off the ground for about seven weeks.

British mining companies have suggested that domestic mining could concentrate their capital at home rather than abroad. It could also help them to solve an awkward tax problem. They now get a credit by way of double taxation relief on foreign earnings – but the credit can only be applied to income arising within the UK, and without such income, the credit does no good.

We have already mentioned the severe land-use conflicts[1–3] likely to arise as low-grade deposits are sought and found in the beautiful, wild, and fragile areas where they are most likely to occur – in the more mountainous areas of older and more resistant rock, areas whose special qualities, like their subterranean contents, are fixed by nature. These conflicts will intensify as such areas and ores become steadily scarcer and more valuable. We shall never have more orebodies or more wild places than we have now, and that is already too little[1].

The haphazard and unjust ways we now have[2] of deciding such conflicts do no credit to our tradition of open government or our ingenuity in improvising new institutions to deal with new social problems. Reforms in the ways in which we now make mineral decisions – or rather, in the absence of policy, let them make themselves – are urgently needed[2]. Without them, important policy options are quietly being foreclosed by actions that seem rational only because discounted cash-flow analysis, not long-range planning, is governing them.

The issues are fundamental: in a rich and overcrowded country that has got that way by importing minerals from all over the world, we must now start to ask whether we can really afford to be so rich, whether our children can afford our being so rich, whether we should instead seek a different sort of richness less costly both to the earth and to our children.

Mining companies are fond of pointing out that they merely give us what we demand – or even, perhaps, what we need, though obviously we are much pressed to demand things we do not need. But it is more complex than that: companies demand things too. In response to what social need did industry provide the hydrogen bomb, the electric toothbrush, Concorde, 400-horsepower automobiles, napalm, fast breeder reactors, the Apollo programme? Partly in response to someone's urge to ensure his twenty per cent after tax?

This may seem unfair to companies which sincerely believe that they are fulfilling a vital social role and that they are being told to mine but then prevented from getting on with it. They are, after all, behaving rationally within the terms of reference of modern, short-term market objectives: and all of us do consume their products. Yet their own interest, just as much as yours, lies in staying in business as long as possible; and doubling mineral consumption every fourteen years is hardly the way to do that.

Mining companies do not yet see this, for most metal depletion lies just beyond their twenty-to-thirty-year time horizons. But eventually they will see it, as the big oil companies have started to do. Perhaps they will then help to work for a sensible resource policy aimed at doing more with less rather than – as now – doing less with more.

The remarkable engineering feats of modern open-pit mining are not yet quite scraping the bottom of the barrel, but they get closer every year. If we keep this new power firmly in check with an understanding and discipline now lacking, we may yet, for many centuries, have a barrel left. Our grandchildren will then have one reason less to ask how rational people could possibly have behaved as we, heedless of their welfare, are behaving now.

Notes

1. See Chapter 3, note 15.
2. See Chapter 3, note 13.
3. See Chapter 3, note 18.

INDEX

This index covers pages xi *to* 100, *including technical terms and all of the Notes. It thus performs the functions of a glossary and of an Index of Authors as well.*

Friends of the Earth and Earth Island

The present environmental crisis is now being recognised as an ultimate and perhaps not very remote threat to man's, and even 'our' planet's, survival. We are finally becoming aware that the land, sea and air on which we depend cannot tolerate the sort of abuse to which we subject them.

Action is required now to develop a life-style based on viable ecological principles, related to the earth's capacity to support the human and other species. But it is one thing to lament (and the capacity for lamentation is one which we humans have developed to a high degree through our history) the tragedy of dying rivers and the curse of industrial dereliction, and quite another to pick an issue, fight the case, and win.

Friends of the Earth Limited is Britain's leading environmental action group. In various translations it today exists in many countries across the world: an international, non-profit-making organisation, with a reputation for political flair and quick, imaginative thinking; a reputation for professionalism, and success.

FOE is light on its feet, academic in its research work, thorough in its strategy, and highly effective in action. Energy is directed into specific campaigns, at both national and local level. More than fifty affiliated FOE groups are now established throughout Britain, dependent on FOE, London, for information, advice, contacts and encouragement.

Campaigns to date (FOE was incorporated in London as recently as 1970) are those against wasteful or excessive packaging (our Government has *promised* us action); the endangered species campaign, aimed at forcing through legislation to ban the importation of the furs, feathers and other by-products of more than sixty animal species threatened with extinction; and the fight to save our National Parks–none is so threatened as Snowdonia–together with other areas of outstanding natural beauty under attack from certain mining companies in their (Government-approved) search for low-grade mineral and other deposits.

Further campaigns at the planning stage are those to tackle Energy (i.e. its waste); Transport; and Sewage.

FOE fights its battles with the help of more than 3,000 members. It survives largely on their financial support. Join us, and beat them!

Friends of the Earth was founded in 1968 by America's leading, longest-serving environmentalist, David Brower. Groups now exist in the United States, United Kingdom, Australia, France, Germany, Holland, Italy, New Zealand, Norway, Sweden, Switzerland, and Yugoslavia.

Earth Island Limited is Britain's leading specialist environmental publishing house, linked with FOE, but like that action group, quite autonomous. Earth Island publishes only those books that *need* to be published. Environmental problems increase by the day. And as the realisation grows that all things are related in delicately-structured eco-systems, so the spectrum of what is properly 'environmental' literature expands. We shall nevertheless go on believing that most books would be best left as trees. . . .

Amory Bloch Lovins, born in 1947 in Washington, DC, transferred to Magdalen College, Oxford, in 1967 after two years at Harvard College. In 1969 he became a Junior Research Fellow of Merton College, Oxford, and in 1971, having taken his MA, resigned his Fellowship to work full-time in London as British Representative of Friends of the Earth Inc. A consultant physicist since 1965, he now concentrates on energy and resource policy.

His recent publications include *Eryri, the Mountains of Longing*, an interpretation and case-study of the Snowdonia National Park with photographs mainly by Philip Evans (FOE/McCall/George Allen & Unwin, 1971/2); *The Stockholm Conference: Only One Earth* (Earth Island, 1972); several technical papers; and evidence for the Joint Committee on Atomic Energy, US Congress (1972), the 'Commission' on Mining and the Environment (1972), and the Committee on Minerals Planning Control (1973).